ELECTRONIC IMAGING TECHNIQUES

a

b

c

d

e

f

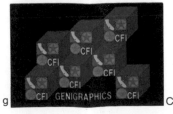

g

C-1. CFI's Genigraphics is an electronic artwork system that creates and processes full-color graphic illustrations with the help of a specially designed computer. There are two ways to initiate activity on the console: existing artwork furnished by the client can be duplicated or altered, or the artist/operator can start with a blank screen and create an original design. He can color, shade, enlarge, or reduce the design—in this case the star (a)—by pressing a few buttons, or he can wipe it off and begin again. To create a two-dimensional effect, a new element may be added to the star—a colored circle (b)—which becomes a background component in the design. By manipulating keys on the console, a spectrum of colors can be tested for compatibility with the foreground image. It is also possible to adjust the relative proportions of the star and the colored circle. To make space for other shapes and elements in the graphic, the star-circle can be shifted from the center to the lower-left-hand corner. If necessary, the image can be cropped, reduced, or magnified to fit into the new area. A graph and lettering can be added to the design (c and d). The effect is quickly visualized and created in minutes at the console. Each new section in the composition can be color- and scale-tested at the time of placement on the screen (e). Any number of images can be isolated, adapted, combined with new formats, and recalled to the screen (f). Most graphic options can be progressively created: in this illustration a three-dimensional effect is shown. Because the equipment can produce visually what an artist mentally "sees," the gap between imagination and reality is reduced—to seconds. When the desired effect is achieved (g), the system's 2,000-line high-resolution film recorder produces permanent hard copy—35mm film or slides. Or the same output can be switched to a U.S. Standard NTSC 525-line television-broadcast signal and recorded on any videotape format.

C-2. Cross-dissolves, 12 exposures.

C-3. Solarization effects.

C-1.

C-2.

C-3.

ELECTRONIC IMAGING TECHNIQUES

A HANDBOOK OF CONVENTIONAL AND COMPUTER-CONTROLLED ANIMATION, OPTICAL, AND EDITING PROCESSES

ELI L. LEVITAN

VAN NOSTRAND REINHOLD COMPANY
New York Cincinnati Toronto London Melbourne

Copyright © 1977 by Litton Educational Publishing, Inc.
Library of Congress Catalog Card Number 76-024376
ISBN 0-442-24771-0

Printed in the United States of America
Designed by Loudan Enterprise

Published in 1977 by Van Nostrand Reinhold Company
A Division of Litton Educational Publishing, Inc.
450 West 33rd Street
New York, NY 10001, U.S.A.

Van Nostrand Reinhold Limited
1410 Birchmount Road
Scarborough, Ontario M1P 2E7, Canada

Van Nostrand Reinhold Australia Pty. Ltd.
17 Queen Street
Mitcham, Victoria 3132, Australia

Van Nostrand Reinhold Company Ltd.
Molly Millars Lane
Wokingham, Berkshire, England

16 15 14 13 12 11 10 9 8 7 6 5 4 3 2 1

Library of Congress Cataloging in Publication Data

Levitan, Eli L
  Electronic imaging techniques.

  Includes index.
  1. Computer animation. 2. Video tape recorders
and recording. 3. Video tapes—Editing. I. Title.
TR897.5.L48   778.5   76-24376
  ISBN 0-442-24771-0

CONTENTS

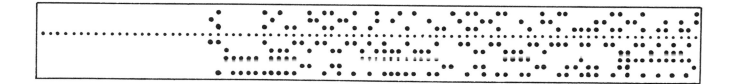

ACKNOWLEDGMENTS

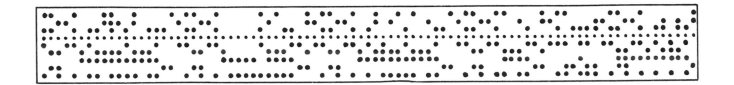

My sincere thanks to the following companies and individuals for their help, information, and illustrative material:

Academic Press, Inc., New York, Azriel Rosenfeld
Animated Productions, Inc., New York, Al Stahl
Bell Telephone Laboratories, Murray Hill, N.J.,
    Dr. Kenneth C. Knowlton, Leon Harmon
BJA Systems, Inc., Willow Grove, Pa.,
    Ralph Weinger
Central Dynamics, Ltd., Montreal, Canada,
    Ken P. Davies
CineMetric, Inc., New York, Larry Plastrik
Cinetron Computer Systems, Inc., Norcross, Ga.,
    Charles A. Vaughn
Colorado Video, Inc., Boulder, Colo., Glen
    Southworth
Computer Image Corp., Denver, Colo., Lee
    Harrison, III
Computer Opticals, Inc., New York, Harvey Plastrik
Consolidated Film Industries, Hollywood, Ca.,
    Mike Sanders, Rex Creekmur
Control Data Corp., New York
Digital Equipment Corp., Maynard, Mass.,
    Stephen A. Kallis, Jr.

Dolphin Productions, Inc., New York, Allan Stanley
Electronic Engineering Co. of California, Santa
    Anna, Ca., George R. Swetland
General Electric Computed Images and Services,
    Syracuse, N.Y., Kenneth R. Anderson
Genigraphics, Ray Christopher
Mathematical Applications Group, Inc., Elmsford,
    N.Y., Dr. Phillip S. Mittelman
National Research Council, Ottawa, Canada,
    Marceli Wein
North American Philips Corp., New York and
    Eindhoven, The Netherlands, Lester Krugman
Oxberry (division of Richmark Camera Service),
    Bronx, N.Y., James Aneshansley
Sony Corp. of America, New York
United Audio Visual Corp., Las Vegas, Nev.

The author is particularly grateful to Dr. Kenneth C. Knowlton of Bell Telephone Laboratories for providing the background information and illustrations for the section on computerized education and research.

PART 1:

CONVENTIONAL PROCESSES

INTRODUCTION

The mind-boggling, faultless computer is, undeniably, one of the world's truly great inventions. Mistakes, unfortunately, are just as plentiful as they were B.C. (before computers). Today, however, the mistakes are nobody's fault.

This book is divided into two sections: the first describes conventional animation, optical, and editing production techniques; the second discusses changing technology in the industry—specifically, the new electronic techniques. The conventional processes, rather than being completely replaced or relegated to a minor role, are actually serving as the foundation for the newer computer-controlled and computer-generated imaging systems. There is now and probably always will be a place for each and room for both. The *old* will complement the *new* and vice versa.

The first part of the book reviews and updates conventional techniques and describes the specialized equipment used by the industry. Established procedures, production processes, and related terminology are explained and defined in language that is as nontechnical as possible. Cross-references help the reader through the more difficult areas. The emphasis throughout the book is on the practical applications of the processes, techniques, and equipment—what makes the systems *go*. The photographs and illustrations complement the text and were carefully selected for their graphic representation of specific processes or techniques.

The second part of the book, dealing with computer-controlled and computer-generated imaging techniques, is as up-to-the-minute as any discussion can be of a field that is still in its infancy. New hardware and new methods are constantly being devised and revised. Electronic imaging has been called dynamic display, motion

1-1. Moving pictures, 1830.

graphics, computer kinetics, etc., etc.—by any other name, indeed! Computer-controlled and computer-generated techniques for producing and recording sequences of motion have actually made it necessary to redefine the generic terms. Commercial production studios and advertising agencies will soon make room and provide special titles for the computer-oriented artist whose training includes higher-level courses in computer programming. At the present time approximately 25,000 students are receiving this training for degrees in motion-picture production, television, and related areas from more than 3,500 faculty members and administrators at more than 800 colleges and universities in the United States.

The potential for the new electronic techniques is great. How great? That question cannot be answered with any degree of authority at this time. Commercial acceptance on a professional level has actually preceded the developmental and experimental stages—an amazing fact in itself. This book is intended primarily for the talented and imaginative people already in animation and commercial-film production and will open the door to more interesting presentations, new techniques, and a variety of unusual illusory effects. It will also be a valuable guide for graphic artists and other students who are interested in a career in the animation and commercial-film industries but are not quite sure of their qualifications. The detailed analysis of the various production techniques will help amateur photographers with only a limited knowledge of the requirements of the industry to attain professional status. Advertising-agency personnel will also find the contents valuable as a guide for additional creativity at both the preproduction and the postproduction levels. As a textbook for schools and colleges it provides clear, detailed explanations and analyses of every phase of production and opens the door to a career in an industry that because of its ever-changing technology will always remain in its infancy.

ANIMATION

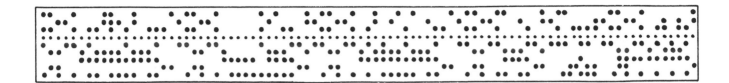

animate (ań-i-māt). vt. 1. to give natural life to; to make alive. 2. to give spirit or vigor to; to stimulate; rouse. 3. to impart an appearance of life to; as a cartoon. 4. to actuate; prompt.

animated cartoon or drawing. n. a series of drawings with slight progressive changes, made and arranged to be photographed and projected like a motion picture.

These definitions from *Webster's New Collegiate Dictionary* state concisely what the terms "animate" and "animated cartoon" mean, but the reader is left in somewhat the same quandary as the animation-studio visitor who, after being shown through the various departments, still had one last question, "Yes, but what makes them move?" This book answers that question as well as many others on the subject.

Although animation is considerably older than other motion-picture techniques, it is the least understood and perhaps the most misunderstood. Many of the production processes that make the animated cartoon "the highlight and happiest part of any visit to local movie houses" are similar to those for a feature film. The same all-purpose technique can be used to produce a wide variety of illusory effects that are unobtainable with other filming techniques. The two words "impossible" and "implausible" are never used in connection with a technique in which the thin line that separates fantasy from reality and actuality from simulation vanishes altogether.

Animation as a motion-picture technique for purposes other than entertainment is usually more effective and certainly more versatile than a live film. Interest is sustained for longer periods of time, and a greater percentage of the subject matter is retained by the viewer. The impact and effectiveness of animation, used by itself or combined with other techniques, are acknowledged through its steadily increasing use by advertising agencies for commercial purposes, by industrial organizations for technical and training purposes, and by educators for teaching any subject.

"Yes, but what makes them move?" The animator makes them move. Also the script writer, the layout man, the background artist, assistant animator, inbetweener, inker, opaquer, checker, animation cameraman, optical cameraman, film editor, sound analyst, sound technician, and a small army of talented and dedicated personnel. Some use pencils; others use pen-and-ink or brushes; still others use cameras, moviolas, sound readers, microphones, tape recorders, and other equipment. The answer to the question "what makes them move" lies in the mountainous pile of drawings prepared under the animator's guidance. Each of the drawings, traced onto a transparent sheet of acetate and opaqued with specific colors, is placed over a rendered background and photographed in sequence by the animation cameraman on successive frames of motion-picture film. The exposed film, with its latent images, is processed, and the illusion of motion is created when the film is projected at a standard rate (24 frames per second). On the screen the individual static drawings are transformed into smooth-flowing, continuous sequences of action. In nontechnical language every member of that small but dedicated army of talented artists and technicians helped make the move possible: the production of an animated film designed to entertain children of all ages as well as their grandparents—and to produce anything from a chuckle to the heartiest belly laugh is a complicated, time-consuming procedure and a very serious business.

Animation was invented by Joseph Antoine Plateau, a Frenchman, who developed the phenakistoscope in 1831. This crudely designed device for showing sequences of motion combined the two features that make modern photography and projection possible. One of the two disks in the device carried the drawings and, mounted on a simple shaft, served as a projector. The drawings were viewed through slits cut into the second disk. The two disks mounted on a common shaft led to the development of the camera shuttle and helped lend credibility to the persistence-of-vision theory. Three years later an Englishman named William George Horner invented a device called a zoetrope, or daedaleum (wheel of life), for showing moving drawings. Two years after the Civil War ended the first American patent for showing "motion pictorially" was issued to William Lincoln. The first animated cartoon was produced by J. Steward Blackton in 1906. The American artist called it *Humorous Phases of a Funny Face*. The development of new procedures, techniques, and equipment paralleled the growth of the animation industry. Each new process had the same basic objectives: to speed production, reduce costs, and eliminate the tedium and monotony associated with many of the conventional procedures.

Following the same general pattern used for the production of motion pictures, an animated film begins with the preparation of a **script**. For animated cartoons as well as television commercials the script is usually developed in the form of captions placed beneath the illustrations, which are arranged in comic-strip fashion. This **storyboard** serves as a guide for the production processes.

The first major step in the production of an animated cartoon is to record the **sound track**. An integral part of any motion picture, the sound track is especially important to an animated production. In the world of fantasy that exists within the borders of the film frame it provides the needed audio background to complement the visual effects and lends credibility to the animated-cartoon characters. Sound tracks recorded after the animation has been photographed are referred to as **postsync** tracks. In most instances, however, the recording of the sound track precedes any of the other production processes, not only with animated films made for entertainment purposes but also with television commercials that include animation sequences. If the recording process takes place after the animation has been photographed (postsync), the film editor is expected to improvise and supply the animator with rough timings as a guide for the animation. These timings indicate the number of frames allotted for a particular action within a scene or for a cartoon character's mouth movements (lip sync). A prerecorded track is preferable and is used whenever possible.

In both instances the original recording on tape is transferred to film (either 33mm Magna-Stripe or 35mm magnetic full-base). The transfer of the sound track from tape to sprocketed film enables the film editor to use motion picture-equipment (moviolas, sound readers, etc.) to synchronize the sound track with the animation (on sprocketed film). The **moviola** is basically a projection device that allows the sound track and the picture portion of the film to be run simultaneously (to **interlock**). The **sound reader**, as the name implies, is used to reproduce and analyze the contents of the sound track. The **synchronizer**, in turn, is an accessory for measuring lengths of film. The results of the film editor's sound-track analysis are entered in the film's blank area alongside the magnetic stripe that

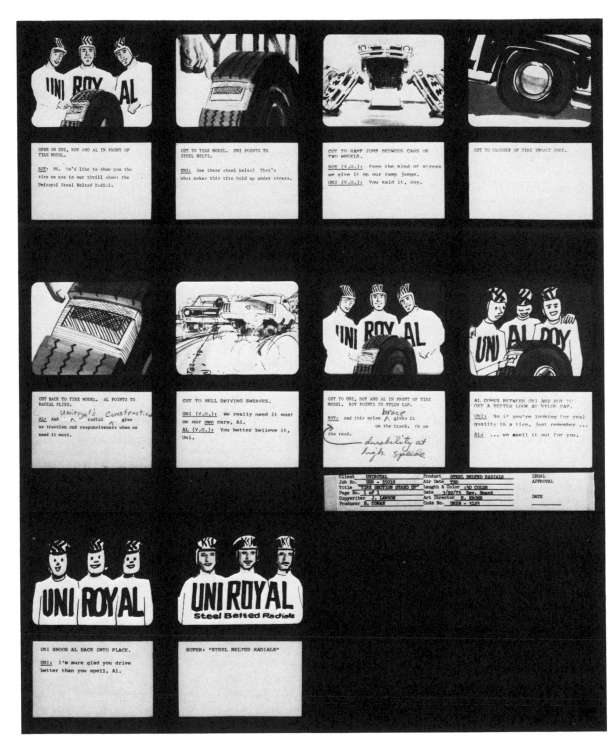

2-1. The storyboard.

carries the sound track. Words are reduced to syllables and rewritten phonetically. Accents and beats are also indicated in the clear area.

The corresponding frame and footage counts are then transferred to **bar**, or **lead sheets**. These sheets serve as a guide for every phase of production, indicating the exact frames in which actions take place along with the timing for each word in the recorded dialogue. Timings for mouth actions, or lip sync, are particularly important, since the attention of the audience is usually focused on the cartoon character's face during exchanges of dialogue. Also included on these sheets—which are actually a visual synopsis of the entire production—are the musical beats and the specific frames in which camera effects such as zooms, fades, and cross-dissolves take place.

The animator, under the supervision of the di-

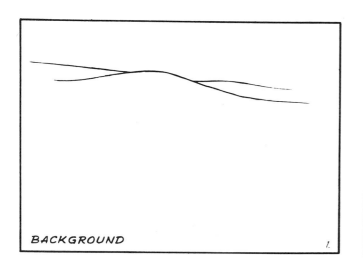

BACKGROUND   1.

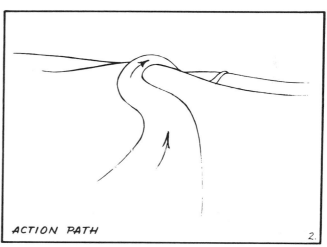

ACTION PATH   2.

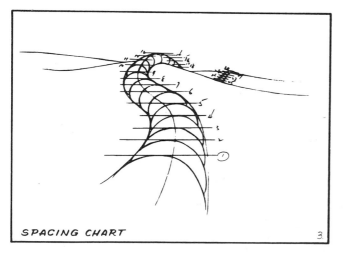

SPACING CHART   3.

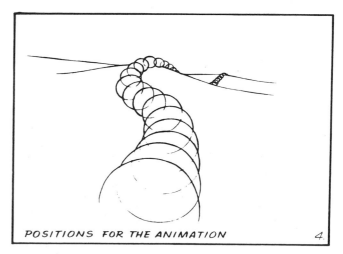

POSITIONS FOR THE ANIMATION   4.

2-2. The animator's layout drawings of an action are based on the film editor's analysis of the sound track. The bar sheets show the number of frame allotted for an action. This information serves as a guide for spacing and timing the animated action.

rector or layout man, does not make every drawing required for a complete action—only the key drawings, referred to as **extremes**. These drawings, registered on pegs in the animator's underlit drawing board, are necessary for showing the action and plotting its continuation. The animator's drawings contain spacing instructions that guide the animator's assistant (**inbetweener**) in preparing the **inbetween** drawings needed to complete the action.

When all of the drawings have been completed and assembled, the action is checked. Flipping the drawings, a process similar to that used in viewing devices at amusement parks, immediately draws attention to flagrant flaws in the action. Another more accurate method for checking the fluidity of the sequence is to photograph the drawings on the animation stand (a camera suspended over a flat surface with a stop-motion motor and provision for registration pegs and underlighting). The process of photographing the animator's drawings is referred to as a **pencil test**. The exposed film, after processing, is viewed on a moviola. If changes in the animation are necessary, the part of the action requiring correction is redrawn.

In the next phase of the production process the drawings are traced on 8 1/2"-×-11" sheets of transparent celluloid or acetate (**cels**) the same size as the drawing paper used by the animator. The **inker** traces each line meticulously with a crow-quill pen in either black ink or another color selected by the animator. After the inker has traced all of the drawings in the scene onto cels, the **opaquer** applies opaque watercolors to the reverse side of each cel in accordance with the instructions included on the model drawings. Opaque colors applied to the reverse side of the cel keep the ink lines from running or smearing and hide the crude appearance of visible brush marks. The inked-and-opaqued cels are in turn checked for numerical continuity and color consistency. In the last of the prefilming processes each cel is matched to doors, windows, and other props rendered on the backgrounds prepared by skilled artists specializing in that area of production. These prefilming checks are in effect a dry run designed to avoid subsequent retakes.

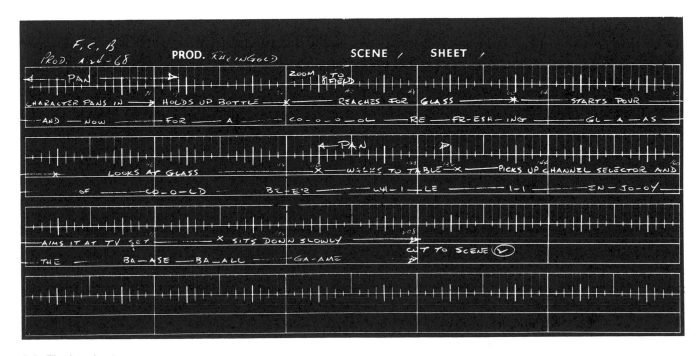

2-3. The bar sheet.

SAME MOUTH ACTIONS CAN FIT DIFFERENT TYPE HEADS

2-4. Lip-synchronization chart. It is not necessary to animate each vowel or consonant in a sentence—in fact it is almost impossible and should not be attempted. The animator should absorb the overall feeling of the dialogue before deciding which words, syllables, or sounds to accent or emphasize and roughly pencil in the key positions. The inbetween drawings are usually sufficient to carry the balance of the animated dialogue.

mouth action	also used for
a	r
e	c
i	l
o	y
u	q
f	s
g	t, z
k	d, h, x
m	—
n	—
p	b, j
v	—
w	—

2-5. These illustrations show how extreme and inbetween drawings are made, with the animator's spacing guides for the inbetweener to follow. In (a) drawings 1 and 5 are the extreme drawings: each of these positions is drawn on a separate sheet of paper, and the spacing guide calls for evenly spaced movements. In (b) the action progressively slows down.

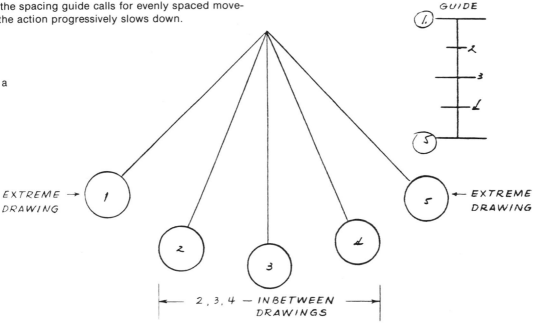

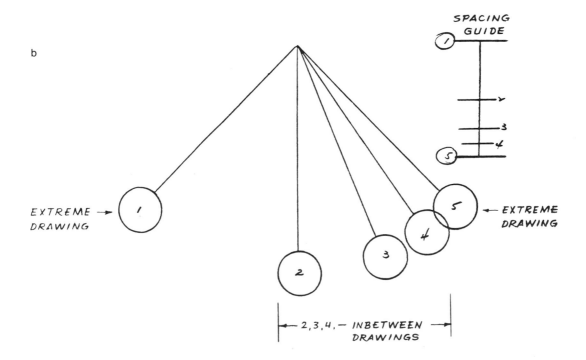

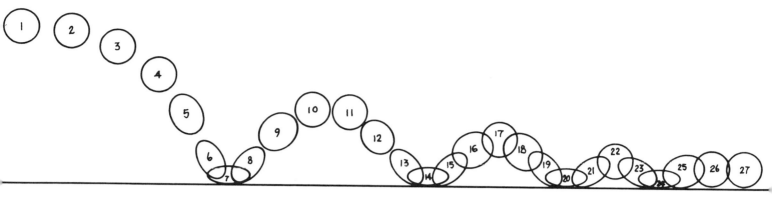

2-6. The action of a bouncing ball illustrates a number of natural physical laws. The apparent squash or flattening of an object in contact with a solid surface and corresponding elongation during acceleration or deceleration are demonstrated, as well as its normal proportions as it moves at a constant speed and its reduction in speed as momentum is lost. Cartoon characters are given a feeling of realism by applying these physical laws.

back

front

2-7. The inked and the opaqued sides of the same cel. The numbers on the corresponding drawings indicate the colors and/or designated shades of gray.

PROD. NO.	CLIENT	PICTURE NAME	NO.	SCENE	FOOTAGE	SHEET NO.	NO. SHEETS
5847	G.E. RADIO	MR. WIMPLE		1	16	1	3

SYNOPSIS: OPENING SCENE — WIMPLE WALKING
ANIMATOR: LEVITAN
ASSISTANT: CRAIG P.
ANIMATION STAND NO. 1
FILM 35 MM - 5V53

DIAL	TRACK	ACTION	DIAL	BKGD.	MOUTH ACTION 1	2	HEAD 3	BODY 4	LEGS (cycle) 5	FLD	PAN	EFFECTS
1			1	SILHOUETTE	BLANK	BLANK	1B	1C	18	12	1	
2	J		2	OF		1A					2	
3	H		3	BUILDINGS		2			19		3	
4			4								4	
5			5						20		5	16 X
6			6								6	FADE IN
7			7		3				21		7	
8			8								8	
9			9		4				22		9	
10	O		0								10	
1			1		3				23		11	
2	O		2								12	
3			3		4				24		13	
4	O	MR. WIMPLE	4								14	
5		IN CYCLE	5			2B	2C	(1) H.U.		15	
6	H	WALK —	6		5	2B	2C		16	
7		HOLDS	7					2		17	
8		RADIO TO	8							18	
9		EAR	9					3		19	
20			0							20	
1			1		4			4		21	
2			2							22	
3			3					5		23	
4			4		3			6		24	
5			5					6		25	
6			6		2					26	
7	T		7		6			7		27	
8	H		8					8		28	
9			9							29	
30			0		4			9		30	
1	I		1					9		31	
2			2		3					32	
3	S		3					10		33	
4			4					11		34	
5			5					11		35	
6			6							36	
7			7		5			12		37	
8	R		8							38	
9	A		9		3			13		39	
40			0							40	
1			1		1			14		41	
2			2							42	
3	D		3		2			15		43	
4	I		4							44	
5			5					16		45	
6			6							46	
7			7					17		47	
8			8							48	
9	O		9		3			18		49	
50			0							50	
1	O		1		4			19		51	
2			2							52	
3			3		1			20		53	
4			4							54	
5			5					21		55	
6			6							56	
7			7					22		57	
8			8							58	
9	S (X)		9		7	BLANK	3C	23		59	
0	T	STARTS	0							60	
1		TO	1		8		4C	24		61	
2	A	SHAKE	2							62	
3		RADIO	3		3		5C	(1) H.U.		63	
4			4							64	
5	A		5		4		6C	2		65	
6			6							66	
7			7				7C	3		67	
8			8							68	
9	T		9		8		8C	4		69	
0			0							70	
1	I		1		4		9C	5		71	
2			2							72	
3	C		3		9		10C	6		73	
4			4							74	
5			5		3	2B	2C	7		75	
6			6							76	
7			7					8		77	
8			8							78	
9			9					9		79	
0			0							80	

Exposure sheets, prepared by the animator, guide the cameraman during the filming process. An **exposure sheet** is basically a visual synopsis that includes frame-by-frame information pertinent to the scene. The animation cameraman places each of the numbered cels on the registration pegs set into the compound table of the animation stand. In some cases as many as four or five cels are positioned over the background, which is registered on a separate set of pegs and can be moved (panned) independently. Panning instructions, along with start and stop positions for effects such as zooms, fades, or cross-dissolves, are also included on the exposure sheets. Zooms and pans must be plotted incrementally before the stop-motion filming begins. The exposed film, along with its latent images, is sent to the laboratory for processing.

If the animation is to be combined with live-action sequences, as is often the case in television-commercial production; if titles are to be superimposed; of it special illusory effects are required, the additions are produced with the optical printer. The **optical printer**, a projection unit operating synchronously with a motion-picture camera, can photograph several lengths of previously processed film simultaneously to produce a composite effect. The preliminary work (photographing the titles, mattes, etc.) needed to achieve the effect, however, is done on the animation stand.

The **dailies** or **rushes** are screened when the processed film is returned, and, if no revisions are needed, the film editor synchronizes picture and sound. In the final stage of production music and effects are mixed with the dialogue track to form a composite track, which in turn is combined with the optical negative to yield the composite print.

Opaqued cels 1, 2, and 3 over their background.

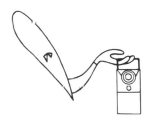

1, drawing for hold cel

2, drawing for hold cel

3, drawing for action cel

←

2-8. The exposure sheet. In the track column the animator writes the dialogue as previously analyzed by the film editor. Under action the animator writes a short description of the general action. The background column shows the kind of background to be used—still or pan. With a pan background the moves are indicated for the cameraman. In columns 1, 2, 3, 4, and 5 the animator indicates the cels that the animation cameraman is to photograph in the proper sequence. The remaining columns on the right-hand side of the exposure sheet are used to supply technical specifications such as cross-dissolves, fades, and zooms.

2-9. A hold cel—a cel that is reexposed for a designated number of frames—is one of the most effective means of avoiding unnecessary inking and opaquing. A hold cel is made if the cartoon character's movement is brought to a complete or partial stop for any length of time. If the character is speaking and only the mouth is moving, for example, the animator can make separate drawings of the head and body and indicate on the exposure sheet that the cel is to be reexposed for a number of frames. The animator need only draw each new mouth action, following the film editor's analysis of the dialogue in the prerecorded sound track.

Certain artists are synonymous with the development of animation as a major motion-picture technique—Walt Disney, Max Fleischer, John Bray, Paul Terry, and Winsor McKay, to name only a few—a medium for amusement, education, scientific research, and simulation. The list is long, but no history of the industry could be complete without mentioning another pioneer, John Oxberry. Although Oxberry began his career as a cartoonist, he was responsible for developing much of the technical equipment that helped the industry advance beyond its infancy.

Although animation is a technique that is sufficient unto itself, imaginative use, especially when combined with other motion-picture techniques, makes possible a wide variety of illusory effects. These effects, in turn, become stepping-stones for the development of new techniques and additional effects. Secure and unchallenged as an entertainment medium, animation has also justified its reputation as a teacher par excellence and, in television, as a supersalesman.

2-11. This film clip from a television commercial produced for the Green Giant Company provides a classic example of the combination of live action with animation. In the first step of the production process the cartoon characters are photographed on the animation stand according to the instructions on the animator's exposure sheets. The same cels are also photographed on high-contrast film, using the animation stand's underlighting unit. The resulting silhouettes on the processed high-contrast film are used as mattes for the subsequent optical combination of the live-action background film with the animation.

→

2-12. The visual-squeeze technique.

2-10. Cyclic action is a shortcut that is used wherever possible in animation production. A complete cycle of action (a walk, for example) is drawn so that the first and last drawings—hookup positions are similar and interchangeable. The cycle is repeated during the filming process for the amount of footage designated by the animator on the exposure sheet.

HOOKUP   HOOKUP

HOOKUP   HOOKUP

HOOKUP   HOOKUP

THE ANIMATION STAND

In all motion pictures the subject "moves." In live action the performer's movements are filmed by the cameraman, but in animation the illusion of movement is achieved through a series of drawings or photographs that are carefully prepared to show progressive stages of an action and photographed in a given sequence by an animation cameraman. To attain the required precision in movement and camera calibration, the cameras and the drawings are mounted on **animation stands**—devices that allow both the camera and the subject matter to move in carefully controlled steps forward or backward, side to side, or both.

The animation stand consists of a motion-picture camera mounted on columns and suspended over a compound table. The camera, driven by a stop-motion motor, allows the cameraman to film individual drawings or cels successively on a frame-by-frame basis. The columns not only support the camera and its motor but also enable the camera to move vertically in relation to the artwork positioned on the registration pegs set in the compound table. The **compound table**, with its ground-glass insert over the underlighting unit for filming transparencies, can be moved in any direction. The vertical camera movement combined with the horizontal compound-table movements make it possible to achieve a wide variety of zoom and pan effects. Previously processed film can also be projected through the camera shuttle onto the compound table; which allows graphic material to be matched up to the film for mattes or rotoscope work. The term "animation stand" is all-inclusive, and referring to the camera, compound table, and control panel.

The animator's stage is the **field guide** which has linear markings arranged to indicate the areas covered by the camera lens. Proportioned to the 1.33:1 aspect ratio of the standard 35mm film frame, each **field size** is indicated by a number corresponding to the actual width of the field in inches: 11F (11 field), for example, is 11" wide and 8 1/4" high; 4F is 4" wide and 3" high. The field guide, used for positioning and composing artwork, is as important to the animation cameraman as the pencil is to the animator. The field guide has punched holes similar to those on drawings, cels, and backgrounds, which enable it to be registered on the peg bars on the compound table.

Zooms, in which the camera moves toward or away from the stationary subject matter registered on the compound table, are probably the most frequently used animation effect. Calibrating a zoom on the animation stand is an involved process. Since the camera is mounted in a fixed position and can only move vertically, horizontal movements are accomplished with the compound table. These horizontal north/south and east/west movements must be plotted in relation to the camera. Four separate and distinct camera and compound-

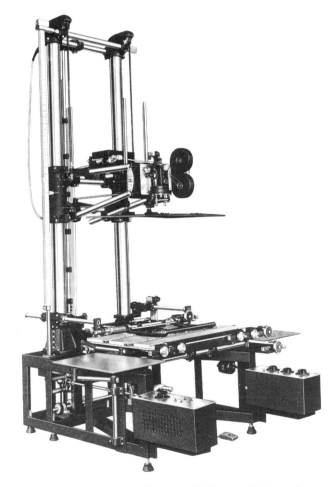

2-13. The animation stand. (Courtesy Oxberry—division of Richmark Camera Service.)

table movements enable the animation cameraman to center the compound table and artwork in any predetermined position. The first of these movements, the **zoom** mechanism, allows the camera to be positioned on any field indicated on the animator's exposure sheets. The position bar on the animation stand indicates the exact **field size** in relation to the artwork. If the artwork is to be photographed on a 12F—a shooting area 12″ wide—for example, the camera is moved to the 12F setting

indicated on the position bar. A counter synchronized with the zoom mechanism permits even greater accuracy of camera position. Follow-focus cams automatically adjust the focus as the camera is positioned on other fields or at intermediate positions. The second and third movements, the **pan** mechanism, allow the artwork to be positioned off-center in relation to the camera. The compound table can be moved to north/south and east/west positions either mechanically or manually, and

zoom counter

traveling peg bar controls and counters

platen

2-14. The compound table. (Courtesy of Oxberry—division of Richmark Camera Service. Photo by Allen B. Howard.)

counters for each movement help the cameraman position the artwork with extreme accuracy. These movements can be calibrated to 1/1000″ if necessary—and in technical animation it very often is. The fourth movement is the **spin** or **rotating** mechanism. It allows the compound table to be spun or revolved independently of the other movements. A counter, synchronized with this spin mechanism, enables the cameraman to position artwork at any angle during the photographic process.

The combinations of these four camera and compound-table movements gives the animation cameraman a great deal of latitude and an amazing amount of control in positioning artwork and creating various zoom, pan, and spin effects. The **pantograph**, a small, tablelike unit attached to the side of the compound table, provides the animation cameraman with a visible check of the camera position in relation to the artwork at any point during the filming process. The pantograph can also be used to plot straight-line or complicated curved pan movements. The counters synchronized with the four compound-table movements ensure an even greater degree of accuracy than the indicating pointer attached to the pantograph.

When all of the movement counters are set at the 0000 reading, the compound table is at the exact center position in relation to the camera. On most animation stands a field guide with a 12″ angular field of view is used as the standard, and each movement from one field to another in an east/west direction is measured in increments of 50 on the east/west counter, which is equal to and indicates 1/2″ of movement. Each full field in the north/south direction is measured in increments of 36. The difference of 14 numbers between the north/south and east/west movements is in direct proportion to the dimensions of the motion-picture frame with its 1.33:1 **aspect ratio**. The counter synchronized with the spin movement has a ratio of 10 numbers to each degree of rotation. One complete table turn in either direction totals 3,600 numbers on the counter. The figures for each of the table movements are reasonably standardized, and slight numerical differences in other animation stands do not require a different method for calibrating and plotting movements from one designated position to another.

To plot a zoom-and-pan movement from one field position to another, the cameraman merely takes readings at both the starting and stopping positions of the movement. To calibrate the moves for each frame of the movement, the cameraman divides the numerical difference between the counter readings at the start and stop positions by the number of frames allotted for the entire effect. It is advisable to begin the zoom-and-pan movement slowly and gradually increase the numerical difference, or spacing. The progressive increase at the beginning of a move is referred to as an **ease-in**, or *taper*; the gradual numerical decrease at the end of the move is known as an **ease-out**. The intermediate moves between the ease-in and the ease-out should be kept at a constant speed.

The pantograph unit is attached to and moves in conjunction with the compound table. The pointer, indicating the camera center in relation to the field guide positioned on the pantograph registration pegs, guides the cameraman during straight-line or complicated curved panning movements. With this panning method counter readings and the subsequent calibrations are unnecessary. The panning path is drawn on animation paper and registered on the pantograph pegs. The spacing for each move on the panning path is determined by dividing the number of moves indicated on the exposure sheets by the length of the path drawn by the animator. By using the north/south and east/west compound-table controls the animation cameraman can move the pantograph pointer along the projected panning path easily and quickly.

Underlighting, or **backlighting**, has many uses. A series of photographic-exposure tests helps to determine the best and most efficient lighting system in relation to the shutter openings on different cameras and animation stands. An opal glass is used for diffusion and even distribution over the light source. The glass is positioned in the opening in the center of the compound table. With underlighting the animator's pencil drawings can be photographed and checked before they are finally inked and opaqued. Mattes for optical printing, the filming of transparencies, and the exposure of graphic material for special effects are all made possible through backlighting. Animation-equipment manufacturers have designed special containers for underlighting purposes, but an ingenious cameraman can quickly and easily make an underlighting setup by wiring several lamp sockets together and testing the system to determine the proper exposure for a given assignment.

Two of the most frequently used effects in motion pictures are the fade and the cross-dissolve. A **fade**, as the name implies, refers to the appearance or disappearance of a scene or image. A **fade-out** is used at the end of a scene or sequence and has an effect of finality. A **fade-in** has the opposite effect and is used to introduce a scene or sequence. A fade is produced by gradually opening or closing the camera shutter for each frame of the effect. Most motion-picture cameras, including cameras adapted for animation photography, have indicators that show the position of the shutter in degrees. These markings serve as a guide for producing fade effects on a frame-by-frame basis by means of stop-

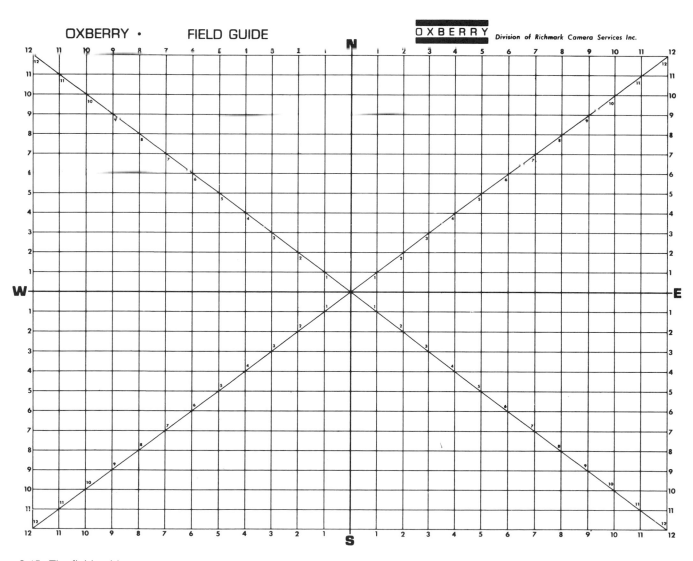

2-15. The field guide.

motion process. One fade-in and one fade-out add up to one **cross-dissolve**, which is used to suggest changes in both time and location. The effect is produced by fading out one scene and fading in the next over the same length of film. Because of the blending or overlapping of picture information, this effect is also referred to as a **lap dissolve**. It is produced by incrementally closing the shutter during the exposure of the fade-out scene for the designated amount of time. With the shutter still in the closed position, the exposed film with the faded-out scene is rewound to the opening frame, and the dissolve is completed by simply fading in the incoming scene over the length of film. This between-scenes transitional effect not only provides the desired visual continuity but is also pleasing to the eye.

Varying degrees of transparency needed for deliberate **double-exposure** effects can also be produced with the camera shutter. Between the two extreme positions, completely open or completely closed, almost any degree of transparency may be achieved. The lower the shutter reading, the more transparent and ghostlike the image becomes. The shutter is also used to produce **burn-in** and **superimposition** effects, in which titles are added to previously exposed film.

To prepare animation used in conjunction with a live-action scene, it is often necessary to position the artwork in relation to the live film. This matching or positioning process is accomplished by projecting frames from the live-action film onto the working surface of the compound table. The pencil tracing of the projected frame from the live-action scene enables the cameraman to position a product, title, matte, or animated action with an amazing degree of accuracy. The same tracing is also used as a layout for preparing the artwork.

This projection process is *not* used to position a simple title or similar subject matter: the **viewfinder**, which facilitates composing and focusing, is adequate. A rack-over viewfinder with ground-glass focusing is mounted on the camera door. A reticle gauge is located at the bottom of the viewing device. The ground glass is precisely mounted and aligned to match the film registration pins in the shuttle. This arrangement permits a previously exposed and processed frame of film to be placed on the fixed registration pins so that the cameraman views first through the film and then through the ground glass, which shows the rectangular-field markings. An optical system rectifies the subject so that the image is right side up and copy can be read normally. The viewfinder has a special control for magnifying the ground-glass markings, allowing extremely accurate positioning.

The camera can also be used as a projector, which makes the **rotoscope** technique possible. This technique is used to duplicate human motion graphically. To produce the effect, live actors, made up and costumed to resemble animated-cartoon

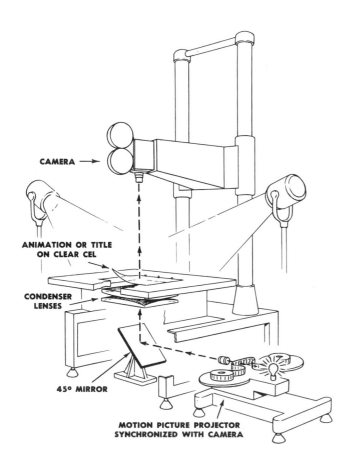

CAMERA →

ANIMATION OR TITLE ON CLEAR CEL

CONDENSER LENSES

45° MIRROR

MOTION PICTURE PROJECTOR SYNCHRONIZED WITH CAMERA

2-16. Aerial-image unit. (Courtesy of Oxberry—division of Richmark Camera Service.)

characters, are photographed in front of neutral backgrounds. The processed film, positioned in the camera shuttle and projected onto the compound table, is traced by the animator. The outlines and details of each figure on the successive frames of film are redrawn on separate sheets of drawing paper, registered to the pegs on the compound table. Each of the drawings is traced on transparent cels and hand-colored (opaqued). In the final stage of production the cels are photographed over appropriate backgrounds. The processed film is synchronized with the corresponding sound track by the film editor.

To meet today's demand for increased animated-film production and lower costs, a variable, high-speed, **stop-motion motor** is necessary in order to handle the tremendous shocks that shutter mechanisms must withstand. It is attached to the animation camera either directly or through an adaptor. The speeds usually range from 60 rpm, or 1/2-second exposure, to 240 rpm, or 1/8-second exposure, enough latitude for slow or fast film stock and for variations in lighting systems. A fast rewind speed operates at 720 rpm, or the equivalent of 1/24-second exposure. The four forward speeds permit a greater choice of f-stops, and the 180 and 240 speeds save time on long, continuous camera runs for titles and similar work.

ANIMATION GLOSSARY

aerial-image photography A unit used to project previously processed lengths of film onto the ground glass positioned between the registration-peg bars on the compound table. Titles and other graphics, inked and opaqued on transparent cels, are placed on the registration pegs and combined with the projected film via stop-motion processes. The overhead-lighting units illuminate the graphics during the filming process. The film in the aerial-image unit is projected on the ground glass that is positioned over the underlighting unit. The aerial head on an optical printer doubles its effectiveness: while running footage passes through the projection head, additional lengths of film, threaded in the aerial head, can be exposed simultaneously to create a wide variety of unusual illusory effects.

amici prism A 90°-angle prism used in conjunction with the conventional lenses on the animation stand for insert shots and tabletop photography.

aniforms A technique (developed by Aniforms) in which two-dimensional cutouts are substituted for the mountainous pile of drawings used in conventional animation. Developed by the famous puppeteer, Morey Bunin, the aniforms, or representations of the characters to be animated, are cut out of a sheet of flexible black plastic material, positioned in front of a black background, and outlined with white paint. Polarity is reversed in the television camera during the filming process. The resultant white image with its black outline provides a unique, line-drawing effect.

animascope A system (patented by Westworld Artist Productions) for producing animated-cartoon effects with live-action filming procedures. The performers are fitted with foam-rubber masks to which a flat makeup is applied and wear specially designed costumes prepared by the art department. The actual filming process takes place on a small stage, and the performers act in front of a special black-velour backdrop. The flat, uniform light falling on the performers does not illuminate any part of the background. Color-reversal stock that provides the highest possible resolution is used for filming in order to produce the matteing elements needed for the optical combination of animation and rendered background. The background is prepared by the studio art department in keeping with the storyboard requirements and is photographed separately on the animation stand.

animated zoom A zoom effect achieved with artwork rather than with camera movement. The graphics for the effect are prepared by duplicating a title, for example, in a number of sizes and positioning each title on a separate cel. The graphics can be moved to any size or position within the film frame.

animation A technique in which the illusion of movement is created by photographing a series of individual drawings on successive frames of film with stop-motion processes. The illusion is produced by projecting the processed film at the standard sound-speed rate of 24 frames per second.

animation board The drawing board used by the animator. Rectangular glass inserts in the center of the board, together with a light source beneath the insert, enable the animator to see the lines on a number of drawings simultaneously. The drawing board is equipped with registration pegs, either fixed or on traveling bars. Movable bars, which are used to position the pan backgrounds and to slide cels for cycle actions, enable the animator to plan actions in relation to moving elements in the scene.

animation camera A motion-picture camera mounted on columns and suspended over a compound table. The camera, driven by a stop-motion motor, films individual drawings or cels successively on a frame-by-frame basis. The columns not only support the camera and the motor but also enable the camera to move vertically in relation to the artwork positioned on the registration pegs set in the compound table. The compound table, composed of a ground-glass insert over an underlighting unit for filming transparencies, can be moved in any direction. The vertical camera movements and the horizontal compound-table movements make possible a wide variety of zoom and pan effects. Previously processed film can be projected through the camera shuttle onto the compound table, permitting graphic material to be matched to the film for mattes or rotoscope work. The **animation stand** includes the complex of camera, compound table, and control panel. See also **matte** in the optical-effects glossary and **rotoscope**.

animation, computerized The generation of motion sequences by means of a computer or the operation of an animation stand by control tapes produced by a computer. In a computerized system the animator generally makes the key drawings, and the computer interpolates the intermediate drawings needed to complete the action. In a computer-controlled animation-recording system photographic instructions are initiated by a tele-type tied to a computer. The punched-paper control tape produced by the programming process directs the animation stand through the entire sequence of movements. See the sections on cinetron and key-frame computer-generated animation in part 2.

animation, multiplane effects A process for producing dimensional effects with cel animation. The camera is mounted on a lathe bed, and the artwork is produced in the conventional manner. The dimensional backgrounds, however, are constructed from a variety of materials and mounted on a movable table. The cels for each exposure are locked in a glass frame positioned in front of the background. The distance between the animation cels and the several background planes creates dimensional effects with continually changing perspectives.

animator The artist or cartoonist responsible for creating the illusion of motion with the drawings produced under his direction. The animator himself does only the key drawings for a motion sequence; the intermediate drawings needed to complete the action are done by the animator's assistant. See also **extreme** and **inbetween**.

anticipation A preliminary action that precedes and emphasizes the main action.

background A realistic or abstract scene rendered with watercolors by the background artist. During the filming process the inked-and-opaqued cels carrying the animation are registered to pegs on the compound table and positioned over the background, which is registered to a separate set of pegs.

back lighting A light unit positioned under the compound table and used to film transparencies, drawings for pencil tests, and mattes for special effects. The light is focused on a rectangular ground glass in the center of the compound table between the top and bottom traveling-peg bars.

bar sheet, or **lead sheet** A visual synopsis of the animation sequence prepared by the animator from the film editor's sound-track analysis. It indicates the number of frames allotted for a specific action, the frames in which dialogue occurs, and the position of musical beats and special effects.

bipack A process utilizing four compartmented film magazines to combine artwork positioned on the compound table and lengths of previously processed film. With this simple optical printer the processed film is loaded into one of the magazine's two feed compartments, and raw stock is loaded in the other. Both lengths of film pass through the camera shuttle simultaneously on their respective routes to each of the two take-up compartments. The artwork is exposed along with the previously processed film to produce composite effects.

blue backing A background used to produce matteing effects. It is exposed to color-blind or non-sensitive film. Subject matter photographed in front of this background produces self-mattes for subsequent optical combination.

breakdown An indication that intermediate drawings are needed to complete the action. The instruction appears on the animator's key drawings along with spacing guides showing how the action should be timed or divided.

bumper footage A specified number of frames exposed at the beginning and end of each scene photographed on the animation stand. The extra footage provides added flexibility for the editor during the postproduction stage.

burn-in (1) The process of superimposing titles over previously exposed film. The subject matter receives additional exposure during the burn-in run on the animation stand. (2) The combination on the optical printer of titles or other graphic material with processed film.

calibrations (1) A marking on the background indicating its position in each frame of the panning action. This information appears on the exposure sheets prepared by the animator and is used as a guide by the cameraman during the filming process. (2) The incremental moves plotted by the cameraman for zoom and pan effects.

camera projection See **rotoscope**.

cel A transparent sheet of celluloid, about 0.005" thick and similar in size to the drawing paper used by the animator, used for tracing and opaquing the cartoon figures. It is punched to fit over the registration pegs on the drawing board and on the compound table.

cel level The position of each cel in the composite picture in relation to the background. To expedite the animation and photographic processes, a cartoon character is often inked and opaqued on a number of cels. If the body, for example, is not animating, it is traced separately on a **hold cel**, which is designated as **bottom level** and positioned closest to the background. The head, in turn, would be inked and opaqued on another cel and placed on the registration pegs over the hold cel. The animating eyes and mouth would be inked and opaqued as a **top-level cel**. With the separation of component parts into top, middle, and bottom cel levels the head and body do not have to be retraced for each frame, and the animation cameraman has to change only the top-level cel for each frame of the animated action.

checker The person responsible for checking the inked-and-opaqued cels for numerical continuity and color consistency. In a dry run that precedes the photographic process the checker places each of the cels over the background and matches the cartoon characters to props if necessary.

collage A montage effect. The subject matter displayed throughout the film frame may consist of a wide variety of graphic elements, including photographs, drawings, or abstract materials. See **montage** in the optical-effects glossary.

color model, or **model drawing** One of several drawings interspersed throughout a scene indicating the colors that are to be applied during the opaquing process. Since many people work on the drawings for a given scene at the same time, the color models assure color consistency.

complete drawing An animator's notation indicating that a drawing should be completed by the inbetweener. If only part of the cartoon character is in motion, for example, the animator would draw the moving part only and add the letters **"c.d."**

compound table See **animation camera**.

continuing zoom A zoom effect in which the subject matter continually increases in size in successive frames of film until it not only fills the frame but continues beyond the field covered by the camera lens.

crawl title The presentation of credits or other information with a panning effect. The lines of copy appear at the bottom of the film frame, pan up through the frame, and disappear over the top.

cross-dissolve (abbreviated **x-dissolve**) One fade-in and one fade-out, probably the most frequently used between-scenes transitional effect. It is intended to suggest changes in both time and location. The effect is produced by fading out one scene and fading in the following scene over the same length of film. Because of the blending or overlapping of picture information the effect is also referred to as a **lap dissolve**. The effect is produced by closing the shutter incrementally during the exposure of the outgoing scene for the designated length of the effect. With the shutter still in its closed position the exposed film with the faded-out scene is rewound to the opening frame. The effect is completed by simply fading in the incoming scene over the same length of film. Cross-dissolves generally provide not only the desired visual continuity but also are quite pleasing to the eye.

cross-fade A transitional effect consisting of two separate cross-dissolves. In the first the outgoing scene is faded to a color or a shade of gray rather than to black. The incoming scene is cross-dissolved from the selected color to complete the effect.

cut The joining of two scenes: one scene ends abruptly and the next scene follows immediately.

cutoff area The outer limits of the film frame that are not visible on television receivers. Titles and other graphic material prepared for television must be confined to the safety field, an area comfortably within TV-transmission limits.

cutout A drawing that is inked and opaqued on thin illustration board, cut out along the outlines, and pasted in place over the background of the scene. Cutouts are used to reduce the number of cel levels.

cycle A series of drawings arranged to form a complete action. A walk cycle, for example, includes the drawings needed to complete one full step and lead in to the start of the second step. The first and last positions of a cycle are referred to as **hookups**.

documentary A film depicting historical, social, or political events, conditions, or people in real-life footage and in an objective, factual manner.

double exposure An effect produced by reexposing the same frame of film, resulting in a ghost image.

effects (abbreviated **effx**) A general term for illusory images such as fades, cross-dissolves, wipes, split-screen shots, montages, matte shots, and multiple-image shots.

exposure (abbreviated **x**) The production of a latent image on film; in animation, the photographic process in which the cels are placed over the background and photographed on the animation stand via stop-motion equipment.

exposure sheet (abbreviated **x-sheet**) A sheet, prepared by the animator and used by the cameraman during filming, that lists each cel, its position over the background (cel level), and the number of exposures it receives and also shows the frames in which zooms and/or pans take place, and position of the background in every frame, and the location of fades and cross-dissolves. It follows the drawings and cells through every stage of development before reaching the animation camera.

extreme A key drawing made by an animator, that illustrates the main action and contains instructions for its continuation. Spacing guides indicating timings and other notations guide the inbetweener in producing the secondary drawings needed to complete the action.

fade An effect used at the beginning or end of a sequence. A **fade-in**, used to introduce a sequence, is produced by gradually opening the camera shutter so that each successive frame of film receives progressively greater amounts of exposure until the density and color values in the scene are identical to those in the original subject matter. A **fade out** is produced by gradually closing the shutter until the last frame of the effect is completely black.

field guide A numbered guide used by the animation cameraman for composing and aligning artwork and for plotting zoom and pan movements. The markings are arranged to indicate the areas covered by the camera lens in use.

field size The area covered by the camera lens: an 11 field (11F), for example, indicates a shooting area that is exactly 11″ wide.

film A thin layer of light-sensitive material over a supporting cellulose base. Rolls of film vary in width from 70mm with 10 perforations on each side of the film frame to 8mm with a single sprocket hole per frame. The perforations match the sprocketed devices in camera, projector, optical-printer, editing, and laboratory-transport mechanisms, providing precise registration at every stage of production. Film expands when warm, contracts when cold, swells when wet, and shrinks when dry. Coating the film base with a gray or jet-black dye will prevent undesirable halation and has no effect on print quality: it merely prevents light from passing through the layer of emulsion and reflecting back from the cellulose base. Films are available in roll form or daylight loading containers and in a wide variety of types, including black-and-white, color, and reversal stocks. Special types are also available, such as film that is sensitive to colors beyond the range of the human eye, film specifically designed for television recording, sound-recording film, high-contrast film for matteing and titles, duplicating stocks for laboratory and optical printing, high-speed rapid-reversal film, and leader stock for editing purposes. Standards outlined by the American Standards Association (ASA exposure indexes) assure a high degree of quality and consistency.

filmograph or **slide-motion film** The use of zoom and pan effects over static artwork to create the illusion of motion—actually a slidefilm with movement.

filmotrip or **slidefilm** A series of photographs, illustrations, and other graphics that is shot in sequence on successive frames of film. During the projection process each frame is visible on the screen for the length of the accompanying narration, which is carried on a disk or tape. A new audiovisual system developed by the 3M Company makes it possible to add sound to the slides themselves. With this sound-on-slide system each slide holder carries its own detachable magnetic sound track. The slides are focused on screens by means of automatic projectors with built-in recording and playback capabilities. To update the subject matter, the sound track is rerecorded and replaced by another detachable sound track. Advertising agencies should find this system quite useful for storyboard presentations.

floating pegs Registration pegs, mounted on a flat metal plate, that can be placed at any point on the animation-stand compound table. They allow the compound table, along with any artwork positioned on it, to be moved without affecting the registration.

gimmick A shortcut such as a cutout devised by the animator to reduce the number of drawings required for an effect. In most cases the use of gimmicks is questionable: any saving in either time or effort for the animator is negated by the additional burdens placed on other production departments.

graphics A generic term for illustrative material, generally not including animation.

hodoscope A display panel containing a number of lights or lamps, each of which may be activated separately. Each lamp flashes for a brief interval corresponding with an action or movement occurring in front of the display panel, which activates the controlling circuits. In some cases the individual lamps may be activated by lengths of projected film focused on the panel. Especially valuable for high-speed, multichannel data recording, a shutterless, fast-framing camera is set up to expose the rapidly changing data that passes in front of the lamps. The technique is especially effective if the projected film that activates the panel carries an animated action sequence.

hold cel A particular cel, group of cels, or background that is exposed for a specified number of frames, as opposed to the cels carrying the action, which are changed in successive frames of film. See **cel levels**.

hookup The drawings for the first and last positions of a movement cycle, which are interchangeable. See **cycle**.

inbetween An intermediate drawing needed to complete an action. The animator makes only the key drawings; the assistant animator, or **inbetweener**, is responsible for the secondary positions. See also **extreme** and **key drawing**.

inker The person responsible for tracing the lines in the animator's drawings on transparent sheets of celluloid.

lap dissolve See **cross-dissolve**.

layout The design of the characters to be animated and/or the plotting of the action.

lead sheet See **bar sheet**.

lily A card divided into spaces containing each of the primary and complementary colors along with a small number of tonal values in the gray scale. It is placed over the last frame of the artwork being photographed and exposed for several frames. After the film is processed, the lily is used for evaluating the colors and tonal values in the print.

limited animation The use of shortcuts to produce the maximum amount of movement with a minimum of drawings. The hold cel is particularly important in this animation technique, since the animator draws only the parts of the figure that are actually involved in action. See **hold cel**.

lip synchronization Matchup of the mouthing of dialogue by an animated-cartoon character with the sound track. The animation is prepared in accordance with the frame counts prepared by the film editor from the sound-track analysis.

matte See **matte** in the optical-effects glossary.

mechanical animation The construction and subsequent photography of three-dimensional models of a mechanism. The subject matter is illustrated by means of directional flow lines, which are inked and opaqued on cels and superimposed on a second camera run. Other movable parts of the mechanism are also rendered on cels and photographed on the compound table with stop-motion processes.

model drawing See **color model**.

model sheet Initial drawings of a cartoon character in a number of representative poses, (made by the layout man or animation director and arranged on a single sheet of drawing paper. Copies are distributed to the animators to serve as guides and assure consistency of proportions, facial expressions, and other details.

multimedia, or **videography** In educational television, a relatively permanent facility available for multiple-image presentations. The programmed material is either channeled to the participating schools via closed-circuit television or displayed on monitor screens by means of various projection devices. Commercial development of techniques that are capable of filling oversized screens with images from a number of sources continues at a brisk pace. Budgets for the production of multimedia presentations, in which the information is transmitted to the screen's beaded surface by as many as six projectors, are proportional to the complexity of the effort and the desired effect. The technique is quite popular with many of the nation's largest corporations and is used for in-plant sales meetings and conventions. Many professional educators have voiced strong opposition to the multimedia presentation, questioning the practicality of a technique that demands a great amount of viewer concentration and requires the absorption of information under abnormal conditions. To these critics the technique is analogous to a reading process in which the reader is asked to scan the contents of a number of pages from several books simultaneously. In a multimedia production the component picture information is usually a joint venture. Contracts for live-film sequences, for example, are assigned to studios with indoor sets, backlot facilities, and filming crews equipped to handle location work; slidefilm presentations, to studios specializing in that type of production; animation, to an animation studio. Each of the separate efforts must be assembled, coordinated, previewed, edited, and synchronized with the corresponding sound tracks—a Herculean task indeed. The use of optical processes, however, eliminates many of the problems associated with the synchronization of composite images. The combined picture information is exposed with matteing processes onto one strand of film that can be threaded into a single projector for normal viewing on conventional screens. Static artwork, slides, photographs, and documents are photographed in different areas of the film frame in accordance with the storyboard requirements. During each of the several runs needed to produce the composite effect the optical cameraman exposes each length of film with its corresponding matte. Live-action and animation sequences are also added optically.

off-register   A vibration effect produced by moving the animation camera in relation to the compound table on successive frames of film. The incremental north/south and east/west movements, in relation to the artwork registered to the pegs on the compound table, create an effect that accents violent animated actions such as explosions or crashes.

ones, twos, threes   The number of exposures that each of the cels in an animated action receives during the photographic process. An action exposed on two frames for each cel, for example, is twice as long and requires half as many drawings as the same sequence photographed on ones. In limited-animation techniques exposure on threes is not uncommon.

opaquer   The person responsible for applying opaque colors to the outlined areas on the cels. He applies opaque watercolors to the reverse side of the cel according to the instructions on the model drawings.

optical printer, optical camera   See **optical printer** in the optical-effects glossary.

overlay   A piece of artwork rendered on thin illustration board and hinged to the background. The overlay and the background sandwich a portion of the inked-and-opaqued cels, creating the illusion of depth.

pan   An effect in which the camera is moved horizontally from one point of the set to another. See also **tilt**.

pan background   A panoramic background designed so that it can be moved in relation to the cels on the registration pegs of the compound table.

pantograph   A small, tablelike unit attached to the side of the compound table. A pointer, indicating the camera center in relation to the field guide positioned on the registration pegs, moves in conjunction with the compound table. The pantograph makes plotting and executing complex compound-table moves a relatively simple procedure.

pegs, registration Fixed pins on the various types of drawing boards used by animators, inbetweeners, inkers, and opaquers. The same peg setup is also used on the compound table for registering drawings, cels, and backgrounds.

pencil test The photographing of an animator's drawings in order to check the fluidity of an action. The drawings are reviewed by screening the processed film on the editor's moviola, and corrections can be made before the drawings are inked and opaqued.

platen A frame-enclosed piece of optically clear glass, which is attached to the compound table to lock the cels in position over the background during the filming process.

postsynchronization See **postsynchronization** in the editing glossary.

presynchronization See **presynchronization** in the editing glossary.

process camera A camera with a stop-motion motor, specifically designed for animation stands and optical printers.

puppet animation A stop-motion technique in which the subject matter consists of three-dimensional figures and backgrounds, which are moved incrementally for each frame of the action sequence.

reading The film editor's analysis of the pre-recorded sound track for synchronization with the animation. The frame counts indicate the number of frames allotted for a specific action and a breakdown of each word of dialogue into syllables.

register The positioning of the animated drawings, the inked-and-opaqued cels, or the background on the registration pegs in the drawing board or the compound table.

repeat The reexposure of an action. See **cycle**.

rock, camera See **off-register**.

rotoscope One of the first techniques for duplicating human motion graphically. To produce the effect, live actors, made up and costumed to resemble animated-cartoon characters, are photographed in front of neutral backgrounds. The processed film, positioned in the animation-camera shuttle and projected onto the compound table, is traced by the animator. The outline and detail of each figure on the successive frames of film are redrawn on separate sheets of drawing paper registered to the pegs in the compound table, and each of the drawings is traced on transparent cels and hand-colored (opaqued). In the final stage of production the cels are photographed over appropriate backgrounds prepared by the art department. The processed film is synchronized with the corresponding sound track by the film editor. The projection process is also used to produce mattes for subsequent optical work: areas within the film frame in which new material is to be inserted are carefully traced by the graphics department, and the mattes are photographed on high-contrast film for optical combination.

run The exposure of film during any part of the photographic process. A double run indicates that the same length of film is exposed twice; two or three runs are often needed to produce multiple-image effects.

scratch-off An operation in which sections of the artwork on successive frames of film are scratched away, removed, or concealed during the photographic process. The camera operates in the reverse mode. When the processed film is projected normally, the artwork appears progressively on successive frames of film.

single-concept film An abbreviated documentary film, usually covering one specific aspect of the subject matter and used as a teaching aid. It is generally distributed in self-threading cartridges, and a continuous loop makes it possible to rerun the film as many times as necessary or to stop at any point for discussion. The film may be run with or without sound.

slidefilm See **filmstrip**.

slide-motion film   See **filmograph**.

sliding cel   One of a group of cels, approximately two or more fields in width, that are used for cycle actions in which the animated-cartoon character pans through the scene.

slit-scan, smear, or **streak photography**   A photographic process in which exposure takes place through a masking slit that is moved in a horizontal plane over the subject matter. Each successive frame of film photographed in this manner provides a time reference relating to a movement, condition, or phenomenon. The technique was originally designed to obtain data pertaining to stress-and-strain conditions, explosive phenomena, and movement patterns normally exposed with high-speed photographic processes. Variations of this masking technique are currently used to achieve unusual abstract-lighting effects in commercial productions. This adaptation requires a motion-picture camera driven by a stop-motion motor. The camera is mounted so that each frame in the sequence of movement—the panning of subject matter behind the slit and the movement of the camera itself—is precisely repeatable. Animation stands with under-lighting are ideally suited for this purpose. During the photographic process a light source directed on the portion of the subject matter that is visible through a slit is moved across one plane of the film frame. During each of the multiple runs needed to achieve the effect the position of the slit in relation to the subject matter is changed incrementally to allow exposure on different planes. The abstract-light patterns in the composite effect create the illusion that the camera is moving through the subject matter along a rainbow-lined route. The technique was developed by Trumbull Film Effects and used in the Stanley Kubrick production 2001: *A Space Odyssey*.

sound reader   A device used by the film editor to reproduce a recorded sound track. The unit is used to analyze the sound track and to provide the animator with frame counts showing the number of frames allotted for an action and/or the frames requiring lip-sync animation. See **reading**.

spin, camera   A spinning effect produced by rotating the animation-stand compound table and the artwork positioned on it.

squash and stretch   Exaggerated changes in the proportions of an animated figure. Squash is a flattening or compressing action that usually results when a moving body strikes a solid surface. Stretch, or elongation, indicates a body traveling at great speed. The exaggerated proportions follow physical principles and lend realism to the cartoon figures.

standard field   The field that is used for most of a studio's filming assignments. It is used for composition, to film titles and other subject matter for optical printing, and to rotoscope previously processed frames of film for matteing. Each studio sets its own standards, based on the equipment in use; in most cases the 12 field is the accepted norm. See **field guide** and **field size**.

still background   A rendered background that remains in a fixed position on the registration pegs of the compound table during the entire photographic process. Compare **pan background**.

stop motion   A photographic process in which the subject matter is exposed on successive frames of film on a frame-by-frame basis rather than continuously, as in standard motion-picture photography. Between exposures the subject matter is moved incrementally to create the illusion of movement.

storyboard   A number of illustrations arranged in comic-strip fashion with appropriate captions. It is an extremely popular method for showing the visual continuity of an animated cartoon or television commercial.

tabletop photography   See **Amici prism**.

take   (1) In animated cartoons and feature-length motion pictures, a reaction indicating surprise. (2) One of several similar shots of a live scene, the best of which is edited into the print.

technamation A stop-motion technique that uses light-polarizing filters to create the illusion of motion. The filters are mounted on a transparent disk positioned between the camera lens and the artwork, usually a large transparency placed over the underlighting unit in the compound table, and rotated incrementally for each exposure. The polarizing axis of each filter changes constantly as it rotates in relation to the other filters. When the axes are parallel, light is transmitted to the artwork; when they are at right angles to each other, little or no light is transmitted. The light source is focused on the ground glass set into the center of the compound table. The transparency is mounted on the registration pegs. Polarizing materials are cut out and fitted into position over each of the action-carrying areas within the transparency. A wide variety of commercially available materials are used to produce an equally great number of motion effects—wheels turning, smoke billowing, or water running, for example. As the filters are rotated and the axes change in relation to the transparency, the density of the applied materials changes continually to create the illusion of motion.

technical animation An animation technique used for educational and training purposes. Instead of animated figures the graphics consist of diagrams, charts, symbols, directional flow lines, and dimensional models of mechanisms with superimposed moving parts. Graphic explanations of intricate mechanisms, demonstrations of new production techniques, or simulations of situations that have no physical counterpart are best left to specialists. The preparation and superimposition of cel animation over a device positioned on the compound table are considered routine by technical animators, who can reduce a complex problem into understandable graphics so that communication with the viewer is established from the first frame to the fade-out at the end of the film.

tilt An effect in which the camera is moved vertically from one position in the set to another. See also **pan**.

trace back A notation on an animator's incomplete drawing indicating to the inker that the missing portions are to be traced from another specified cel within the scene.

traveling pegs Movable bars, set into tracks on the animation-stand compound table, that are equipped with registration pegs for positioning sliding cels and pan backgrounds.

videography See **multimedia**.

visual squeeze An animation technique in which the subject matter consists of short animation sequences, still photographs, abstract forms, and a wide variety of graphics designed and arranged to produce the illusion of movement. An unusual amount of visual information can be compressed within the 1,440 frames of film in a 1-minute television commercial.

wipe effect The replacement of one scene by another by means of matteing processes. The mattes, actually wipe patterns, are animated transitional effects that are photographed on high-contrast film on the animation stand. See also **wipe effect** in the optical-effects glossary.

zoom An effect in which stationary subject matter within the shooting field is made to appear progressively larger (**closeup**) or smaller (**longshot**) in successive frames of film by moving the camera closer to or further away from it. See also **calibration**.

OPTICAL EFFECTS

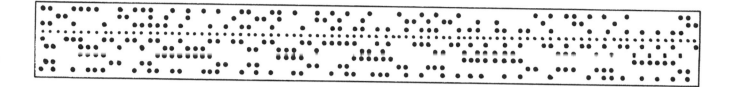

In the completion stage of production all of the loose ends are tied together by the film editor into a neat package, consisting usually of a number of 1,000' film cans. The package is rushed to the optical-services department, and the contents of the several reels are carefully reviewed by the film editor and the optical-layout man. The editor's instructions, indicated by grease-pencil markings on the work print, are usually quite explicit. In addition to frame and footage counts for each scene that is to be duplicated, the **work print** contains notations that specifically identify areas in which transitional effects take place. The location of matte shots, insert material, freeze frames, double framing, reverse actions, and other effects are also pinpointed by small strips of white tape to punctuate the editor's instructions. The grease-pencil entries, translated into filming instructions on optical-layout sheets, serve as a guide for the optical cameraman in producing the **answer print**, which summarizes the talents of the film editor and the optical-services department, whose primary objective is a film presentation that is accurate in terms of visual continuity and color rendition.

Transitions are visual effects used to blend a series of related scenes into a harmonious whole, providing a smooth flow of picture information. They improve the visual continuity by eliminating choppy abrupt scene cuts and separate the pro-fessionally produced film from the amateur offering. Some transitional effects are custom-designed to create specific moods in harmony with the film's subject matter. Producers with slimmer budgets must be content with the more traditional between-scenes effects such as cross-dissolves and wipes. Fade effects are used either to introduce a sequence or to signal its ending.

The **fade—-in** or **-out**—is the most frequently used effect, the easiest to produce, and the least expensive in the optical studio's price list. A fade begins or ends with a completely black frame of film. In a fade-in picture detail becomes progressively more distinguishable until all of the colors and tonal values are duplicated in the final frame of the effect. The technique is the same regardless of the length of the effect: the shutter is opened incrementally, increasing the amount of light passing through the lens to produce proportional increases in exposure. In a fade-out, with its attendant feeling of finality, the picture receives proportionate decreases in exposure as the shutter is incrementally closed until the picture is completely blacked out. Fade effects can be done on the animation stand or with the optical printer: in either case the shutter is either opened or closed by degrees. Most motion-picture cameras can create fade effects during the actual filming process, but in-camera fades reduce the film editor's flexibility during the postproduction stage, and the savings in time and cost by avoiding opticals are negligible. Fade effects can be produced by several other methods: by reducing the lens aperture, by gradually dimming studio lights, and by rotating polarizing filters positioned directly in front of the lens mount. Fading glasses and lengths of film in which the tonal values range from clear through opaque black offer additional alternatives, but none of these other methods approaches the efficiency of the variable shutter.

One fade-in and one fade-out add up to one **cross-dissolve**, the most frequently used between-scenes transitional effect. It is intended to suggest changes in both time and location. The effect is produced by fading out one scene and fading in the second scene over the same length of film. Because of the blending or overlapping of picture information the effect is also called a **lap dissolve**. It is produced on the animation stand or with the optical printer by closing the shutter incrementally during exposure of the outgoing scene for the designated length. With the shutter still in the closed position the exposed film with the faded-out scene is rewound to the opening frame, and the incoming scene is simply faded over the same length of film. The effect not only provides the desired visual continuity but is also quite pleasing to the eye.

Mattes—opaque masks—are used to block out areas within a frame so that picture information on another length of film can subsequently be inserted. Mattes, which are actually opaque silhouettes, conform in size, shape, and position with corresponding areas on the film that serve as the background for the picture information, on separate lengths of film, that will subsequently be inserted in the matted, unexposed areas. Masking the inset area within the background film is only a temporary measure in producing the effect. In the first step of the optical process the positive matte, on high-contrast film, is exposed with the background film. The high-contrast negative matte, in the second of the two optical-camera runs needed to achieve the effect, is exposed with the film that carries the picture information to be inserted in the matted area. Unlike the **static** mattes just described, **traveling** mattes involve a series of mattes, on successive frames of film, that correspond with constantly changing matteing areas in the background film. In either case the basic matteing process makes possible the combination of animation with live action, the superimposition of titles, and the creation of a wide variety of unusual visual effects—without the risk of undesirable double exposures.

Unlike the cross-dissolve, in which the images in each scene are blended and more or less visible throughout the entire effect, the picture information in both scenes of a **wipe** effect is visible but separated by a definite line of demarcation between the incoming and outgoing scenes. The shape of the line varies with the wipe pattern used to achieve the effect. **Wipe patterns**, animated and photographed on high-constrast film, serve as mattes and conceal progressively larger areas of the outgoing scene on successive frames of film. These same areas are progressively revealed during the second camera run until the incoming scene is visible in its entirety. Unlike the cross-dissolve, in which the exposure varies throughout the length of the effect, the exposure during a wipe effect remains constant at 100% during each of the two camera runs. The negative and positive high-contrast mattes that are used to produce the effect are carefully cataloged and identified by number in **wipe charts** used by the film editor and optical cameraman. Depending upon the wipe pattern, the transitional effect not only adds visual interest but also suggests changes in both time and location. Two types of wipe effects are available to the filmmaker. The **hard-edge** wipe, in which a visible line separates the areas between the incoming and outgoing scenes, is used most frequently. The line of demarcation between the incoming and outgoing scenes is nonexistent in the **soft-edge** wipe: instead of a visible line the scenes are separated by a narrow, gray, out-of-focus area that conforms with the leading edge of the wipe pattern in use. In either case the effect is produced by exposing the positive pattern, on high-contrast film, with the outgoing scene for the designated length of the effect—the patterns are available in various lengths. In the second of the two camera runs the film is rewound to the opening frame with the camera shutter in the closed position. The high-contrast negative pattern is then exposed with the incoming scene.

Both camera runs receive 100% exposure. Except for the lines of demarcation between incoming and outgoing scenes or the narrow, gray, out-of-focus areas in soft-edge wipes, the high-contrast positive and negative wipe patterns that act as mattes are never visible to the viewer. If they were placed over each other in the optical printer's projection shuttle, the leading edges of the two lengths of film would match precisely and result in a completely black, opaque frame.

A wide variety of unusual illusory effects can be produced with the optical printer and used either as transitions or to create specific moods. The **out-of-focus** effect, for example, can be used for either purpose. As a transition the outgoing scene is simply thrown out of focus progressively in successive frames of film until the picture information is blurred beyond recognition. The camera shutter remains completely open at all times during the single run needed to produce the effect. If, for example, 80 frames are allotted for the total effect, the outgoing scene is gradually thrown out of focus for the first 40 frames, and the incoming scene is then exposed, beginning with a completely blurred

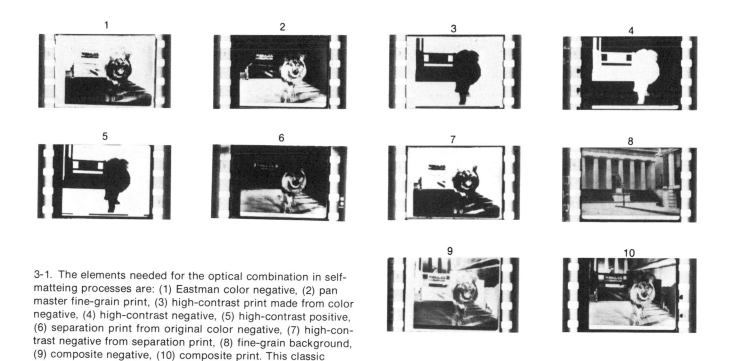

3-1. The elements needed for the optical combination in self-matteing processes are: (1) Eastman color negative, (2) pan master fine-grain print, (3) high-contrast print made from color negative, (4) high-contrast negative, (5) high-contrast positive, (6) separation print from original color negative, (7) high-contrast negative from separation print, (8) fine-grain background, (9) composite negative, (10) composite print. This classic example of the self-matteing process was produced by the MPO studio in New York for the Dreyfus Fund. Filmed under the watchful eye of its trainer, the lion was photographed on a studio stage, not in the city's financial district. During the optical combination of the various film elements that made the effect possible the high-contrast positive matte of the lion (5) was photographed with the background film of Wall Street (8). On the following camera run the lion (2) was combined with the negative high-contrast matte (4) to produce the composite effect (10).

STRAIGHT ACROSS AND DIAGONAL PATTERNS

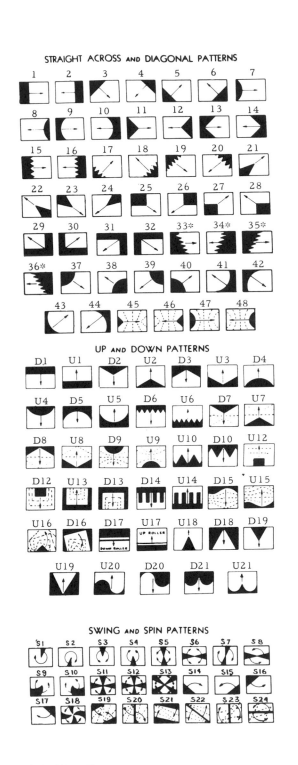

UP AND DOWN PATTERNS

SWING AND SPIN PATTERNS

3-2. Wipe effects.

EXPANDING AND CONTRACTING PATTERNS

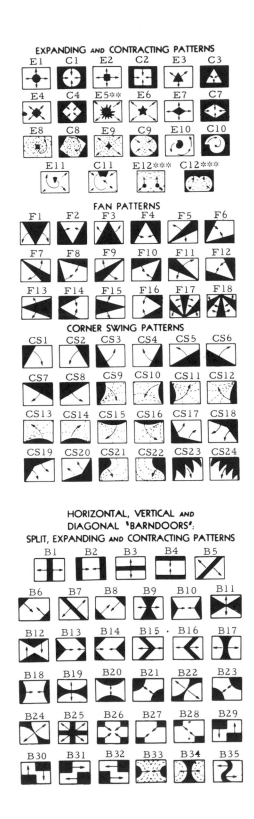

FAN PATTERNS

CORNER SWING PATTERNS

HORIZONTAL, VERTICAL AND DIAGONAL "BARNDOORS": SPLIT, EXPANDING AND CONTRACTING PATTERNS

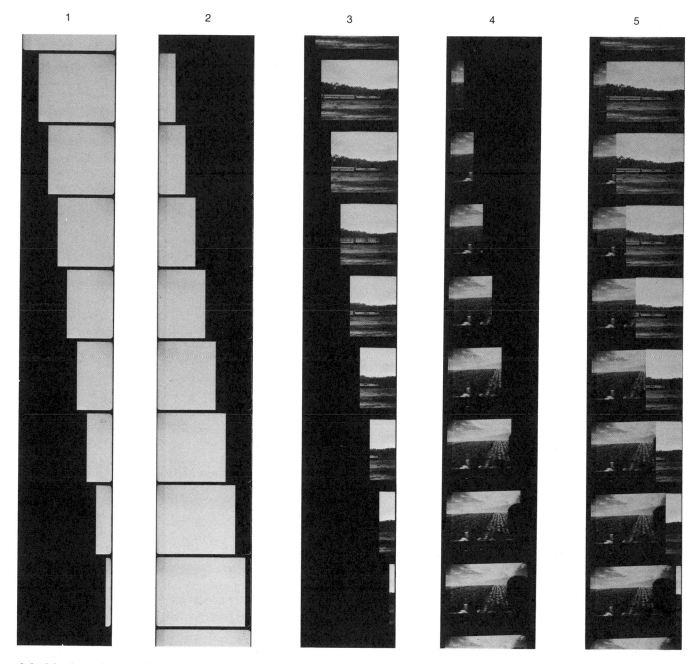

3-3. A horizontal-wipe effect, eight frames: (1) the high-contrast matte for wiping the outgoing scene, (2) the high-contrast matte for the incoming scene, (3) the first optical camera run combining the matte with the outgoing scene, (4) the second optical camera run combining the matte with the incoming scene, (5) the completed effect in the processed print.

frame, for the remaining 40 frames. The picture information is in completely sharp focus by the last frame. The out-of-focus optical effect is relatively inexpensive and can be used to introduce dream sequences, to create horror effects, or to form abstract backgrounds for superimpositions.

Optical effects in which the picture information is deliberately distorted or diffused can also be used to create transitions or to establish a mood. The production of **distortion** and **diffusion** effects for transitional purposes is quite similar to that of out-of-focus effects. The main difference between the two lies in the special materials needed to produce the effects, which may include almost any substance that can be positioned in front of a lens and that allows some unobstructed light to pass through while scattering the remaining light in many directions. The diffusion, usually unpredictable, produces areas of softness and sharpness within the same frame. The effect is used to create an air of mystery, an eerie atmosphere, or a feeling of vagueness. The materials used to achieve diffusion effects range from commercially available glass disks with patterned surfaces to loosely woven fabrics and small pieces of metal window screening. Lenses that either elongate or flatten images in any direction are used to produce distortion effects. In most cases the materials used to produce these effects are supported by frames positioned in front of the camera lens. The degree of diffusion or distortion can be controlled to some degree by exposing the film being duplicated with stop-motion processes. Incremental changes in the position of the materials in front of the lens are made for each exposure. As in the out-of-focus effect, the images in the last few frames of the outgoing scene and the first few frames of the incoming scene are usually so distorted or diffused that the scenes can be spliced as a direct cut.

Many illusory effects can be produced by optically duping **library footage** over the selected scenes. Smoke, fire, fog, or lightning, for example, can be used to create situations that might be too hazardous for direct in-camera photography.

Reversing action is another effect that has withstood the test of time in the motion-picture industry. Originally introduced to elicit chuckles from the most sophisticated audiences, it is now used to produce a wide variety of effects for almost as many purposes. In practically all cases the effect is produced with direct in-camera photographic processes or with the optical printer. With live photography the film is run through the camera with the shutter in the closed position for the length of time needed for the effect. With the camera motor reversed and the shutter reopened, the scene is photographed normally, and the last frame exposed is at the beginning of the roll in the processed film. The process is even simpler with the optical printer: the normally photographed and processed scene is merely rewound, threaded on the projection side of the optical printer, and duplicated on the raw stock in the optical camera. When it is projected normally, the last frame of the action is the first frame visible on the screen—the action is reversed.

The optical printer can also be used to add or drop any number of designated frames from a sequence of action in order to speed up or slow down the sequence. A single frame can be exposed repeatedly on successive frames for any specified length of film footage in order to **freeze** an action within the scene being duplicated.

Even the most sophisticated optical-printing equipment can do little to improve the quality of the image in the footage being duplicated. It is impossible, for example, to restore detail in shadow areas that are almost completely transparent. Conversely, the optical-services studio can add clouds to a clear sky, lighten dark areas by selective printing, or alter contrast by reexposing a scene on a more suitable positive or negative duplicating film or through exposure correction and special developing gammas. The printer cannot perform miracles—no optical effect can cover a bad cut. The producer who hopes that "the editing will save it" or "the music will save it" or "the opticals will save it" probably has a film that isn't worth saving. In many cases an unusual optical effect requires more planning than preparation. Similarly, cost and effectiveness do not necessarily go hand in hand. More often than not the most effective optical for a specific situation is also the least expensive to produce. The effects possible with the optical printer are indeed limited only by the filmmaker's imagination.

THE OPTICAL PRINTER

To creative personnel in the motion-picture industry all special, illusory, and transitional effects are associated with the optical-electrical-mechanical device known as the **optical printer**. The variety of effects that can be achieved with this versatile filming unit is limited only by the collective imagination of the directors, editors, and technicians who are aware of the printer's capabilities and include them in the preplanning stages of production. The optical printer, one of the most important technical developments in film, is a projection unit that operates synchronously with a motion-picture camera. It is designed to duplicate the picture information in several lengths of previously processed film and at the same time to add the optical and transitional effects that are needed to produce the composite image. The print derived from the optical negative provides the optimum visual continuity envisioned by the writers and directors in the preliminary stages of production. Technological advancements have steadily increased the printer's capabilities so that filming assignments that once required a considerable amount of planning and preparation are now regarded as routine. The new technology in turn has initiated a reassessment of conventional filming procedures and resulted in great savings in time and a significant reduction in postproduction costs.

For operation the lengths of previously processed film that are to be duplicated, along with other film elements needed to produce the composite effects (mattes, wipes, inserts), are threaded in open-end magazines on the projection side of the printer. The specially designed projection shuttle, which is synchronized with the shuttle and transport mechanism on the camera side of the printer, can accommodate the combined thickness of several lengths of film simultaneously with no loss in registration. As it travels through the projection shuttle, each frame of processed film that is to be duplicated on the raw stock in the optical camera is illuminated by a light source located in the projector's lamphouse. A shutter in the controlling light valve maintains proper exposure in relation to the various film planes. The lamphouse can be fitted with color-correction filters, which enable the layout man and the film editor to match and reproduce specific colors.

Matteing processes are used to eliminate the possibility of undesirable double exposures during the optical combination of several lengths of film. Mattes, or masks, prepared by the studio's art department, are designed to block out specific areas of the film being duplicated. They are photographed on high-contrast film on the animation stand and exposed with the previously processed lengths of film. On subsequent camera runs picture information on other lengths of film is inserted in the previously matted, unexposed areas. The matteing process makes it possible to combine animation with live action, add titles, superimpose product shots, or create an unusual variety of illusory effects.

Many effects that are difficult to achieve with conventional photographic processes are accomplished easily and precisely on the optical printer. Transitional effects such as fades and cross-dissolves, for example, can be produced quite easily during the filming process itself, although built-in effects do restrict the film editor and remove much of the flexibility that is so desirable during the postproduction stages. Fades and cross-dissolves of any length from 4 to 128 frames can be programmed with the printer's dissolve mechanism. Transitional wipes are produced by exposing the desired wipe pattern (a type of matte) with the footage from both the incoming and outgoing scenes on successive camera runs. Flips and spins, depending upon their function, can also be regarded as transitional effects. Entire scenes or individual frames from incoming and outgoing scenes may be used. Flipped frames produce a mirror image, reversing right and left, an important consideration in scene composition—any printed matter within the flipped frames will read backwards.

The optical printer is certainly a versatile filming tool and in many ways duplicates or exceeds the best efforts of the studio darkroom technician. It can enlarge or reduce the contents of entire scenes or portions of individual frames with little effort; it can freeze- or skip-frame an action, slowing down one part of a scene while speeding up the balance; it can reverse an action. In-camera optical effects achieved with multiple runs—reexposing the same length of film several times—are best done with the optical printer, since the ever-present possibility of error and consequent necessity of reshooting the subject matter negate any saving in either time or cost. Included in this category of effects are double exposures, multiple images, montages, split-screen shots, superimpositions, and inserts achieved with either static or traveling mattes.

Designed and engineered to meet the demands and requirements of the most inventive and imaginative producers, optical printers incorporate features that help meet deadlines and lower budgets. Newly designed models feature modes of operation and controls designed to ease the cameraman's task and reduce the possibility of error.

3-4. Model 1600 special-effects optical printer. (Courtesy of Oxberry—division of Richmark Camera Service.) The elements are: (1) aerial-projector tilt, (2) aerial reduction and blowup (16mm:35mm), (3) aerial-image variable-speed autofocus, (4) variable soft-focus control, (5) automatic iris, (6) camera tilt, (7) accepts 400', 1,000', 1,200', and 400' bipack magazines, (8) independent high-performance servomotors for camera and projectors, (9) digital eletronic frame and foot-frame counters, (10) camera follow-focus for reduction and blowup (Super-8:35mm), (11) digital electronic frame and foot-frame totalizer counters for camera, (12) circujt-breaker panel, (13) camera-projector and aerial-projector controls, (14) skip-add cycler, (15) speed-control module with digital speed readout, push-button control, continuously variable control dial, and fast rewind, (16) film take-up tension control, (17) predetermined-frame counter.

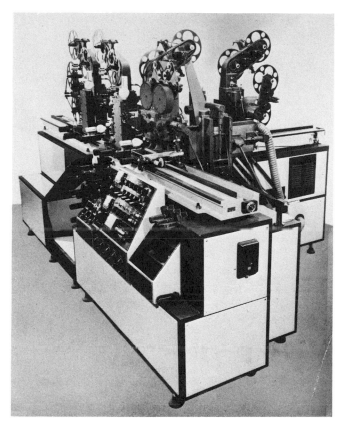

3-5. Modular optical printer. (Courtesy of Oxberry—division of Richmark Camera Service.)

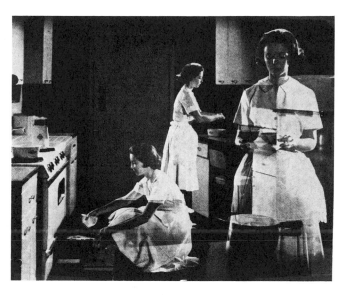

3-6. This deliberate double-exposure effect was produced for a Corning Glass Works television commercial. It required a considerable amount of planning: the background was photographed separately on one length of film; on another stage each of the housewives was photographed in front of a black background. All of the film elements were combined optically to produce the composite effect.

3-7. These film clips from a television commercial produced for the Brillo Manufacturing Company are excellent examples of the multiple-exposure effect. A variation of the split-screen effect, the multiple images are produced with preopticals and mattes. The composite effect is achieved by repositioning the processed preoptical film elements during several camera runs.

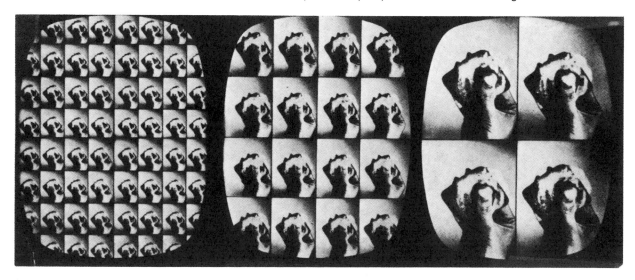

3-8. The split-screen effect.

A choice of filming speeds, for example, makes it possible for the cameraman to select the most convenient speed for a specific assignment. The cameraman also has at his fingertips a number of automatic controls, including skip-frame and multi-print mechanisms, that remove much of the tedium from the filming process. At the push of a button in the printer's control console the cameraman can add or drop a designated number of frames from a sequence of action. The sequence is exposed from the processed film on the projection side of the printer and rephotographed to the required length on the raw stock in the camera.

The condenser lenses that illuminated lengths of film in earlier printers have been replaced by light systems in air-cooled lamphouses. The new achromatic-lens systems provide up to 50% more illumination on the film passing through the projection shuttles, and they are corrected for color as well as for spherical and coma aberrations. Dichroic filters mounted on pivots in front of the lamp source make it possible to introduce subtle variations in color temperature. In addition to the conventional lenses wide-screen processes—which have aspect ratios greater than the standard 1.33:1 format—require anamorphic and other lens types designed to squeeze images during filming and to restore them to normal proportions during projection. Zoom effects from 3:1 enlargments to 5:1 reductions, optically centered or off-center and with pan and tilt movements from one area within the film frame to another, can be achieved quickly, accurately, and with a minimum of preliminary plotting and calibration. Automatic follow-focus systems, coaxially constructed for maximum response and stability, assure image sharpness at every point in the zoom movement.

Most animation stands and optical printers feature interchangeable film-transport mechanisms. Removing a 16mm shuttle-and-sprocket assembly and substituting a 35mm unit take only a few minutes; realigning optical centers takes even less time. The changeover capability makes reductions and enlargements from one film width to another a relatively simple task and increases the versatility of the printer. Recognizing the needs of the low-budget producer, some manufacturers also provide for Super 8mm and Super 16mm interchangeability.

Newly designed optical printers with four projector-head systems allow precise control of all film elements required for matte photography. Each of the four projector heads carries either the insert material, the background film, or the male and female high-contrast mattes, which allows individual positioning of the film elements, an important consideration since mattes must match precisely within 0.00002″ of each other in order to avoid the undesirable black halo around the insert. Matte carriers that advance each frame of film through a projection shuttle in stop-motion fashion provide yet another control.

Other features that make the optical printer the most versatile of filming tools are superimposition devices used to align titles in relation to the background film in the projector head, specialized lens attachments for producing vertical and horizontal spin effects, mechanisms that permit the operator to straighten or reposition a scene by tilting the picture information, and viewfinders that enable the optical cameraman to preview an action in a dry run and make appropriate improvements in composition. Also included are a number of automatic controls, including a buckle switch that immediately stops the camera and activates a warning light to indicate a film jam or other camera malfunction. These devices not only improve operational efficiency but also assure precision at every stage of the filming process, resulting in a greatly improved end product.

Analogous to the exposure sheets used by the animation cameraman, the optical cameraman works with **layout sheets** that indicate frame and footage counts and the exact order and length of scenes, effects, and transitions. The frame counts contained in these guides correspond to the frame and footage counts indicated on the film editor's work print. Frame and footage counters on the projection side of the printer as well as on the camera side provide a visible readout and check during every step of the filming process. Other counters indicate the north/south and east/west positions of the camera and other related controls. The **liquid gate**, or **wet gate**, a relatively new accessory, is a self-contained unit that removes or minimizes the

effect of scratches, dirt, dust, and surface abrasion during the optical-printing process. The film being duplicated is totally immersed immediately before exposure in an **aquarium**, a housing unit that contains perchlorethylene, a liquid with a specific refractive-index effect on film that makes the salvage process possible. In operation the liquid in which the film is immersed fills the abrasions, eliminating light refractions that are intensified as a result of contrast buildup.

3-9. The liquid gate. (Courtesy of Oxberry—division of Richmark Camera Service.)

The optical-electrical-mechanical device known as the optical printer has justifiably been described as one of the most important technical achievements of the motion-picture industry. Since at least one effect that cannot be produced with direct in-camera photography is included in practically every film production from the full-length feature film to the 1-minute commercial, it is obvious that most of the film seen on television receivers or in movie houses has passed through the projection shuttle of the optical printer. The author was once asked during an interview about the optical printer and its place in the industry. The answer is as applicable today as it was then: "Cameras take pictures. Optical printers complete them."

OPTICAL-EFFECTS GLOSSARY

achromatic lens A lens system used to duplicate previously processed lengths of film. Filtering the printing light through the lens eliminates chromatic aberration and the resultant color fringing.

aerial-image head An accessory on the projection side of the optical printer that doubles its effectiveness. In addition to the running footage that passes through the projection head other lengths of film threaded in the aerial head can be exposed simultaneously to create a wide variety of unusual illusory effects.

afterimage See **tail**.

anamorphic print A print in which the aspect ratio is greater than the 1:33:1 standard. Squeezed or compressed by means of an anamorphic-lens system on standard-width film during the filming or optical-printing process, the picture information is restored to its normal proportions during projection by complementary deanamorphic lenses.

answer print See **optical print**.

bipack See **blpack** in the animation glossary.

blowup The duplication of picture information on 16mm film to the 35mm width. In most cases the results are unsatisfactory because the grain in the 16mm film is magnified proportionately.

burn-in See **burn-in** in the animation glossary.

china girl A relatively short length of film containing a medium closeup of a model whose skin tones, clothes, and background include all of the primary and complementary colors. Also included within each of the film frames are color patches that are used as a guide by the timer for making optical or laboratory color corrections.

Cinex A series of filters positioned within windows cut into a circular disk. The device, mounted in front of the optical-camera lens, is used to expose representative frames in a scene before the actual production of the composite. The frames of processed film are evaluated by the film editor and the optical timer, who select the appropriate correction filters for the optical-printing process.

contact printing The production of a negative or positive (depending upon the polarity of the film) by running the original film in direct contact with raw stock past the printing light.

cording An identification technique in which short lengths of string or tape are attached to the perforations at the beginning and end of one of the scenes wound on a reel so that selected lengths of film can be located and made in to fine-grains or dupes.

cross-dissolve See **cross-dissolve** in the animation glossary.

cross-fade See **cross-fade** in the animation glossary.

cue sheet See **cue sheet** in the editing glossary.

cut The abrupt end of a scene or action. The following scene is spliced at the cut point.

diffusion and distortion effects Optical effects in which the picture information is deliberately distorted or diffused. Used as a transitional effect or to establish a mood, the procedures for producing these effects are quite similar to those used for producing out-of-focus effects. The main difference between the two lies in the special materials that are needed to achieve the different effects. Special materials for diffusion include almost any substance that can be positioned in front of a lens and that allows some light to pass through unobstructed while scattering the remaining light in many directions. The diffusion, usually unpredictable, produces areas of softness and sharpness within the same film frame. The effect is used to create an air of mystery, an eerie atmosphere, or a feeling of vagueness. The materials used to achieve the effects range from commercially available glass disks with patterned surfaces to loosely woven fabrics and small pieces of metal window screening. Lenses that either elongate or flatten images in any direction are used to produce distortion effects. In most cases the special materials are supported by frames positioned in front of the camera lens. The degree of diffusion or distortion can be controlled to some degree by exposing the film being duplicated with stop-motion processes. Incremental changes in the position of the materials in front of the lens are made for each exposure. As in the out-of-focus effect, the images in the last few frames of the outgoing scene and the first few frames of the incoming scene are usually so distorted or diffused that the two scenes can be spliced as in a direct cut.

double printing The combination of picture information from two lengths of negative film to produce a composite print.

dropped shadow An optical-printing process designed to improve the legibility of titles. To produce the effect, the black title, on a length of high-contrast film, is exposed with the background film to produce the shadow. The white title, on another length of high-contrast film, is offset during the burn-in run to complete the effect. The offset title is usually positioned above and to the left of the black title.

duping The reproduction of picture information from one length of film on a second length.

duplicate color negative See **internegative**.

effect A generic term for illusory effects of any kind.

fade See **fade** in the animation glossary.

fine-grain A film with a finer grain than that in most raw stock. Fine-grain film is used for optical printing and for duplicating work; resultant prints have a finer texture.

flash frame A deliberately overexposed frame of film that serves as a cue mark for the editor or optical cameraman.

flash pan, or **zip pan** A transitional effect in which the camera is panned away from the subject matter. The speed of the panning action and the resultant blur, during which detail is not distinguishable, is used to suggest a change in location.

flip A transitional effect in which a scene seems to rotate on a center axis until the image becomes a thin line—actually the thickness of the film itself. The incoming scene develops from this thin line into a full-frame image to complete the effect. In practice the outgoing scene is threaded into a flip device positioned between the printer lamphouse and the camera lens. During the filming process the cameraman rotates the flip device frame by frame for the length of the effect. Halfway through the effect the incoming scene replaces the outgoing scene in the flip device, and the stop-motion rotation continues until the scene, in the final frame of the effect, is in the normal projection position.

freeze frame Reexposure of the picture information in one frame of film on successive frames, for the specified length of the effect, in order to stop an action within the scene being duplicated.

grading The scene-to-scene exposure balance throughout a film presentation.

high-contrast film A contrasty film used for photographing titles, mattes, and other elements that are combined with other footage during the optical-printing process.

internegative A negative produced from an original color-reversal print. Negatives derived from print material are referred to as **duplicate color negatives**.

interpositive A denser-than-normal print used for color optical or duplicating work. It has a distinguishing orange-colored base or mask.

jump cut See **jump cut** in the editing glossary.

lap dissolve See **cross-dissolve** in the animation glossary.

layout The timings and frame counts in the film editor's work print, which are transferred to layout sheets for the optical cameraman. The layouts specify the length of scenes to be duplicated, the frame in which transitionals take place, the location of superimpositions, matte runs, and other effects.

liquid gate, or **wet gate** A self-contained unit that removes or minimizes the effect of scratches, dirt, dust, and surface abrasion during the optical-printing process. In operation the film being duplicated is totally immersed immediately before exposure in an aquarium, a housing unit that contains a liquid whose refractive-index effect on film makes the salvage process possible. The liquid in which the film is immersed fills the abrasions, eliminating the light refractions that are intensified by contrast buildup.

master positive Fine-grain print film used to produce a duplicate negative.

mattes An opaque silhouette that conforms in size, shape, and position with a corresponding area on the film, which serves as the background for picture information on separate lengths of film that will subsequently be inserted in the matted, unexposed areas. The masking of the inset area within the background film is only a temporary measure in the achievement of the effect. In the first step of the optical process the positive matte, on high-contrast film, is exposed with the background film. The high-contrast negative matte, in the second of the two optical camera runs needed to complete the effect, is exposed with the film that carries the picture information that is to be inserted in the matted area. Unlike the **static** matte just described, a **traveling** matte consists of a series of mattes on successive frames of film, corresponding with the continually changing matteing areas in the background film. In either case the matteing process makes possible the combination of animation with live-action scenes, the superimposition of titles, and the creation of a wide variety of unusual illusory effects without undesirable double exposures.

montage An effect in which a number of related images are visible within the film frame simultaneously. It is used in many ways for as many different purposes. In a feature-length production the film editor may use a series of extremely short lengths of film spliced together to create an impression. There is no attempt at visual continuity: the effect is used to save time, film footage, and several pages of explanatory script. In a television commercial montage is not only a necessity but a merchandising aid as well. In a 1-minute commercial montage is used to increase the amount of information reaching the consumer. Compare **collage** and **visual squeeze** in the animation glossary.

multiple-image shot An optical effect produced by duplicating the picture information within the film frame as many times and in as many positions as the film editor specifies.

optical effect The duplication and/or addition of picture information on the optical printer: any effect that is produced by optical means rather than with direct, in-camera photographic processes.

optical print, or **answer print** A print produced on the optical printer; it contains the duplicated scenes to which superimpositions, transitional effects, matte inserts, or special effects have been added and is complete with respect to visual continuity. The approved answer print becomes the guide for the production of the **release print** (see the editing glossary).

optical printer, or **optical camera**   An optical, electrical, and mechanical device that combines the functions of the camera and the projector and is used to duplicate the picture information in a number of lengths of previously processed film simultaneously and to create transitional and special effects.

out-of-focus effect   An out-of-focus effect can be used either as a transition or to create a specific mood. As a transition the outgoing scene is simply thrown out of focus progressively in successive frames of film until the picture information is blurred beyond recognition. The camera shutter remains completely open at all times during the single run needed to produce the effect. If, for example, 80 frames are allotted for the total effect, the outgoing scene is gradually thrown out of focus for the first 40 frames, and the incoming scene is then exposed, beginning with a completely blurred frame, for the other 40 frames, ending with completely sharp focus. This relatively inexpensive optical can be used to introduce dream sequences, create horror effects, or form abstract backgrounds for superimpositions.

probe   A device used to measure and check the intensity of the light source in an optical printer.

process camera   See **process camera** in the animation glossary.

projection printing, or **optical printing**   The process of duplicating film by projecting it through an optical system similar to that in an optical printer. The projected film is exposed on the raw stock in the printer's camera.

projection shuttle   The shuttle through which lengths of previously processed film are projected.

It is capable of accommodating several thicknesses of film simultaneously, and the projected images are exposed on the raw stock in the printer's camera.

reduction print   The duplication of picture information on 35mm film in the narrower 16mm width. The size of the grain is not as important in reduction as in enlargement. See **blowup**.

reversal original   A positive picture image produced from in-camera raw stock by reversal processing.

reversal processing   A laboratory process that produces a direct positive image from the in-camera film instead of the usual negatives.

reversing action   A pioneering motion-picture technique, originally introduced to elicit chuckles from the most sophisticated audiences and now used to produce a wide variety of effects for almost as many purposes. In practically all cases the effect is produced with direct, in-camera photographic processes or with the optical printer. To photograph live action, the film is first run through the camera with the shutter closed for the required length of time; then, with the camera motor reversed and the shutter reopened, the scene is photographed normally. When the film is processed, the last frame exposed is at the beginning of the roll, and the action is reversed when the film is projected. With the optical printer the process is even simpler: the normally photographed and processed scene is merely rewound, threaded on the projection side of the optical printer, and duplicated on the raw stock in the optical camera. The film is projected normally to reverse the action.

rotoscope   See **rotoscope** in the animation glossary.

run   See **run** in the animation glossary.

scene   An action or incident, filmed in an appropriate location or setting, that forms a part of a related sequence in a film presentation.

separation positive A black-and-white print produced from a color negative. Each of the primary colors in the negative is reproduced through the appropriate filter in the corresponding shades of gray on a separate length of film. The lengths of film are used to produce a color-corrected negative for printing purposes.

sequence See **sequence** in the editing glossary.

shot A specific scene and its exposure.

skip frame An optical effect, used to speed up an action, in which designated frames are omitted during printing.

solarization A polarity-reversal effect produced by exposing an image to intense light for long periods of time. Imaginative directors and graphic artists use variations of the technique to produce unusual illusory effects. In practice solarization effects are achieved with mattes derived from the live-action color film. In the first step of the process the live-action footage is duplicated on a **pan master** (see the editing glossary). Its gradations from black through intermediate shades of gray to white are used to produce the high-contrast mattes needed to achieve the effect. During the optical combination the negative and print films are projected onto color film through selected filters to produce psychedelic effects.

split-screen shot A shot in which the film frame is divided into a number of sections, each of which contains different picture information. Mattes and multiple runs are used to mask each part of the film frame in turn and to add the insert information in the previously matted areas. A hidden split, in which no definite line of demarcation between the separate picture areas is visible, is achieved by offsetting the wipe patterns on high-contrast film. The invisible-split effect is produced by overlapping the leading edges of the wipes that are used as mattes, resulting in a narrow, gray, out-of-focus area.

start mark, or **sync mark** See **start mark** in the editing glossary.

step-contact printing The reproduction of picture information from a length of previously processed negative or positive film that runs in direct contact with the raw stock. Depending upon the polarity of the film used for printing, either a positive or a negative image is produced. Each frame of film, traveling intermittently through the transport mechanism, is exposed to the printing light in turn.

superimposition See **burn-in**.

tail, or **afterimage** An undesirable trailing or secondary image caused by improper phasing of the camera shutter and the transport mechanism.

take See **take** (2) in the animation glossary and **outtake** in the editing glossary.

title Verbal information relating to the accompanying picture information. In the industry's infancy captions were substituted for dialogue, and titles can be used to establish a mood or reflect the film's subject matter both in full-length feature films and in 1-minute television commercials; the variety of treatments is unlimited. See also **burn-in**.

transitional effect An effect, such as a cross-dissolve or wipe, designed to create a smooth flow of visual continuity between scenes. A fade is similar to a transitional effect but is generally used at the beginning or end of a sequence.

traveling matte See **matte**.

ultrasonic cleaner A device in which film is cleaned by ultrasonic energy. The solvent residue resulting from the cavitation, or boiling, effect as the film moves through the cleaning solution is removed by a forced-air, flash dryoff.

variable-speed motor A drive used on animation and optical cameras that enables the cameraman to select a convenient filming speed by simply turning a controlling dial. The camera can deliver full power at all speeds and operate on a stop-motion, frame-by-frame basis.

wet gate   See **liquid gate**.

wipe   An optical effect in which one scene is gradually replaced by another in the same length of film. Unlike a cross-dissolve, in which the images in each scene are blended and more or less visible throughout the entire effect, the picture information in both scenes of a wipe is visible but separated by a definite line of demarcation. The shape of the line varies with the **wipe pattern** used to achieve the effect. Wipe patterns, animated and then photographed on high-contrast film, serve as mattes and conceal progressively larger areas of the outgoing scene on successive frames of film. These same areas are progressively revealed during the second camera run until the incoming scene is visible in its entirety. Again unlike a cross-dissolve, in which the exposure varies throughout the length of the effect, the exposure during a wipe remains constant at 100% during each of the two camera runs. The negative and positive high-contrast mattes that are used to produce the effect are carefully cataloged and identified by number in **wipe charts** used by the film editor and the optical cameraman. Depending upon the wipe pattern, the transitional effect not only adds visual interest but also suggests changes in both time and location. There are two types of wipe effects available to the filmmaker: in the more common **hard-edge wipe** a visible line separates the areas between the incoming and outgoing scenes; in a **soft-edge wipe** the line of demarcation is nonexistent. Instead of a visible line the scenes are separated by a narrow, gray, out-of-focus area conforming with the leading edge of the wipe pattern. In either case the effect is produced by first exposing the positive wipe pattern on high-contrast film together with the outgoing scene for the designated length of the effect (wipe patterns are available in various lengths) and then, in the second camera run, rewinding the film to the opening with the camera shutter closed and exposing the high-contrast negative-wipe pattern with the incoming scene. Both camera runs receive 100% exposure. Unlike the hard-edge wipe, in which the leading edges of the negative and positive wipe patterns are precisely aligned during each of the two camera runs, the soft-edge wipe is produced by slightly overlapping the wipe patterns. This offset produces the soft, out-of-focus effect that separates the outgoing and incoming scenes. Except for the lines of demarcation or the out-of-focus areas, the wipe patterns are never visible to the viewer. If they were placed over each other in the optical printer's projection shuttle, the leading edges of the two lengths of film would match precisely and result in a completely black, opaque frame of film.

zip pan   See **flash pan**.

PROCESSING

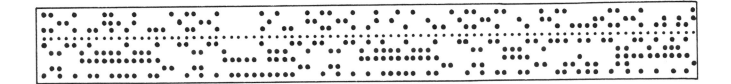

Processing is the series of continuous, automatic laboratory operations in which the latent images exposed on film are rendered visible by treating the film with chemical solutions. Guided through a series of precisely controlled baths, the previously exposed film is developed, rinsed, fixed, and rewashed.

In the first stage the exposed film is spliced to other lengths of sprocketed film. This seemingly endless celluloid chain travels over rollers that guide the film through the developing solution. The latent images are rendered visible by the chemical bath, washed as the film passes through a second bath, fixed in a third bath, and rewashed before entering the drying chamber. At this point the tonal values in the negative are exactly opposite to those found in the original subject matter: blacks, for example, appear as clear areas on the developed film, and the polarity of the whites and light areas is completely reversed. The developed negative (the film originally exposed in the camera) is rewound on plastic cores and used to produce the positive image.

In the next stage of the printing process the emulsion side of the developed negative and the emulsion side of the raw stock used for printing are placed in direct contact with each other. In this position light passing through the developed-negative film exposes the printing film. The pro-cedure used to obtain the negative image on the in-camera film is repeated for the print film. After passing through the developing solution, it is guided through a rinse bath, in which the developing-chemical residue is removed, fixed, and rewashed before entering the drying chamber. It is rewound on positive cores and returned to the production studio along with the negative roll of film.

Specialized emulsions require different types of chemical solutions. The more contrasty film types needed for optical mattes and titles require developing agents that have a higher reactivity than the chemical solutions used for low-contrast results in black-and-white-negative film emulsions. Reversal-film stocks, by comparison, require special handling. With this type of in-camera raw stock the developed film yields a direct positive image—there is no negative. The film is processed to produce a negative image and reprocessed. During this second phase the developed silver on the in-camera film is bleached out, and the film is reexposed and redeveloped to yield a positive image. The fixing bath is used only after the second of the two developing procedures has produced the positive images on the print film.

In a more complicated procedure known as **coupler development**, the values in color film are reproduced to duplicate those found in the original subject matter. The developing process begins when the exposed silver halide forms metallic silver. The oxidized developing agent reacts with the film's chemical couplers to produce insoluble dyes, which are proportional to the amounts of metallic silver formed during the developing process. These dyes are the means for reproducing the original color values. As in black-and-white-negative or reversal-film processing, rinses or wash baths are used to remove residue chemicals from previous solutions. The color values of the negative images are complementary to those in the original subject matter.

The direct positive color images obtained from color-reversal film are also produced by the coupler components. If the couplers exist in the raw-stock emulsion layers, the original in-camera film is developed in a single chemical solution; if the couplers are extracted from the developing solution, each of the film's three emulsion layers is developed separately. The basic processing procedure is the same in both cases. In the first phase of the process the in-camera color-reversal film is developed in a black-and-white bath, which produces a negative image. The unused silver halide in the film's emulsion layers is then reexposed, and the chemical action of the couplers produces positive images that duplicate the original color values. This description is an oversimplification, since the exposed in-camera film passes through more than a dozen different developing, coupling, bleaching, fixing, and washing solutions before entering the drying chamber in the final step of the long and involved processing procedure. Corrections, if needed, are limited to overall color-balance adjustments rather than to individual color changes, which are possible only with color-negative film.

In addition to the more or less standard types of developing agents manufacturers produce a wide variety of chemicals for motion-picture-laboratory use, including, for example, reducers, used to decrease density; intensifiers, which produce the opposite effect; preservatives, which stabilize the dyes in color-print film; hardeners, which prevent scratches and abrasions in print material; prebath chemicals, used before the actual developing process begins; and paraffin and carbon tetrachloride solutions, used to coat the sprocket-hole areas, eliminating the annoying clicking sound that would otherwise accompany the projection process.

In addition to processing the large number of specialized films used for as many different purposes, the laboratory technician is often an unsung, behind-the-scenes hero. Making latent images visible on print film is, by current standards, only a small part of the many services performed by the motion-picture laboratory. The effects of scratches and surface abrasions can be eliminated or reduced by immersing the film in a **wet gate** during the printing process, which, coupled with the efficient service that allows the behind-schedule producer to meet impossible deadlines, cannot be measured in dollars and cents.

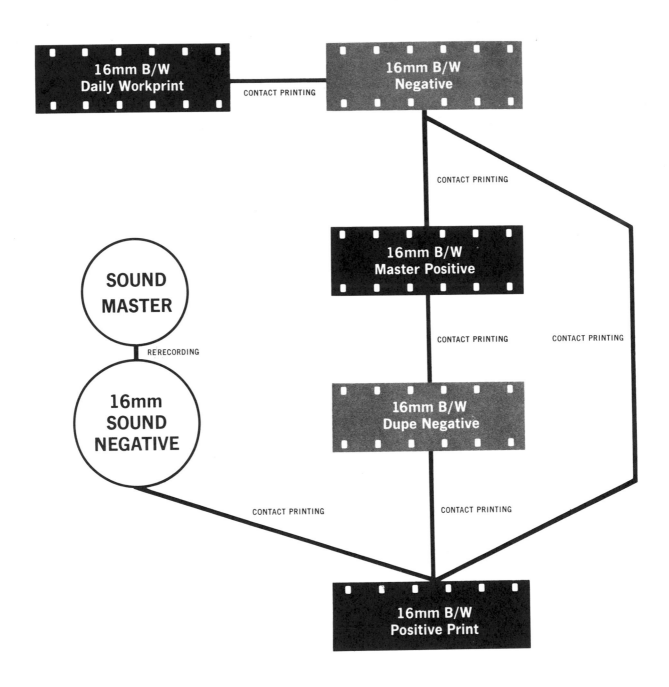

4-1. Printing-flow chart—16mm black-and-white negative.
(Courtesy of General Film Laboratories.)

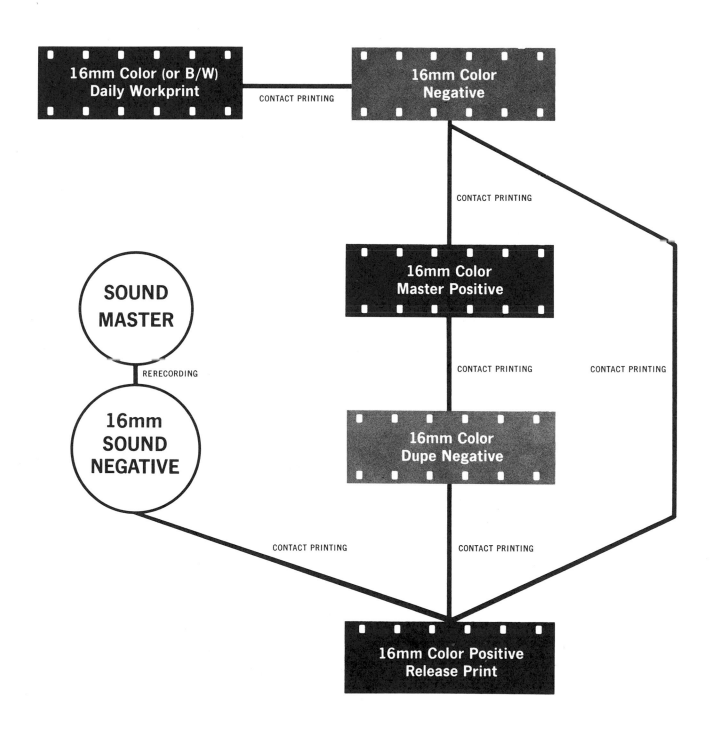

4-2. Printing-flow chart—16mm color negative. (Courtesy of General Film Laboratories.)

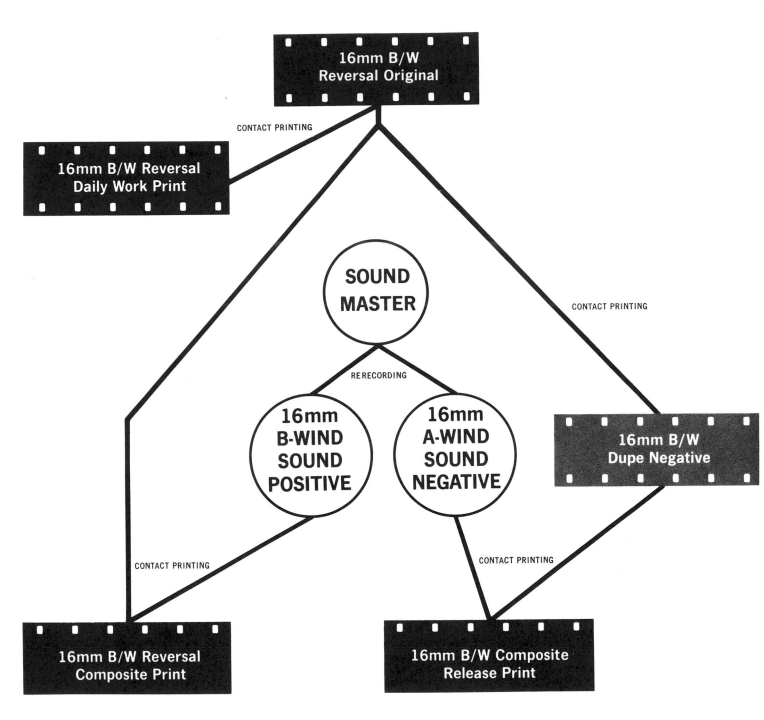

4-3. Printing-flow chart—16mm black-and-white reversal original. (Courtesy of General Film Laboratories.)

4-4. Printing-flow chart—16mm color reversal original. (Courtesy of General Film Laboratories.)

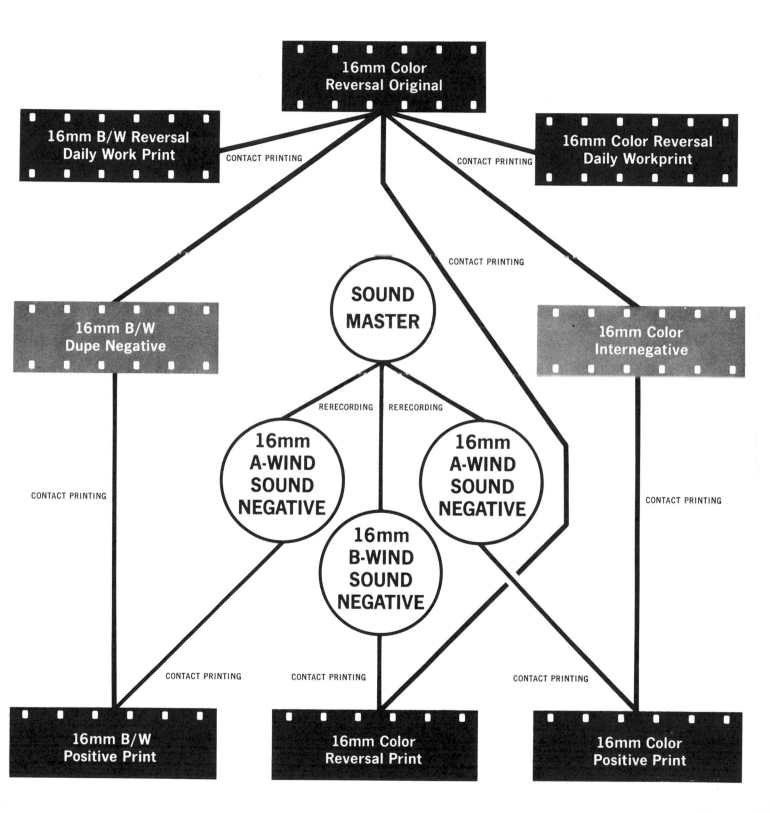

16mm Color Reversal Original

16mm B/W Reversal Daily Work Print

16mm Color Reversal Daily Workprint

CONTACT PRINTING

CONTACT PRINTING

CONTACT PRINTING

SOUND MASTER

16mm B/W Dupe Negative

16mm Color Internegative

RERECORDING

RERECORDING

16mm A-WIND SOUND NEGATIVE

16mm A-WIND SOUND NEGATIVE

16mm B-WIND SOUND NEGATIVE

CONTACT PRINTING

CONTACT PRINTING

CONTACT PRINTING

CONTACT PRINTING

CONTACT PRINTING

16mm B/W Positive Print

16mm Color Reversal Print

16mm Color Positive Print

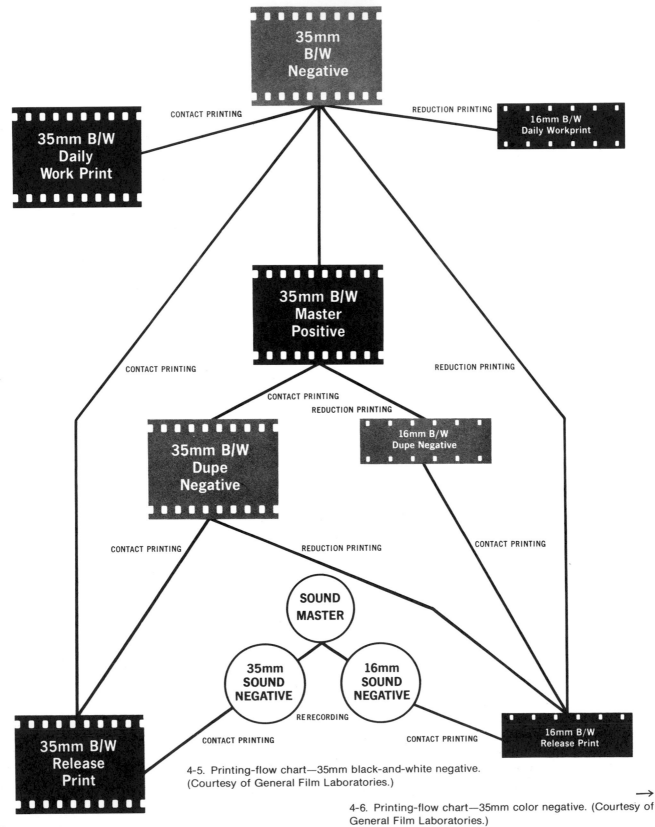

4-5. Printing-flow chart—35mm black-and-white negative. (Courtesy of General Film Laboratories.)

4-6. Printing-flow chart—35mm color negative. (Courtesy of General Film Laboratories.)

→

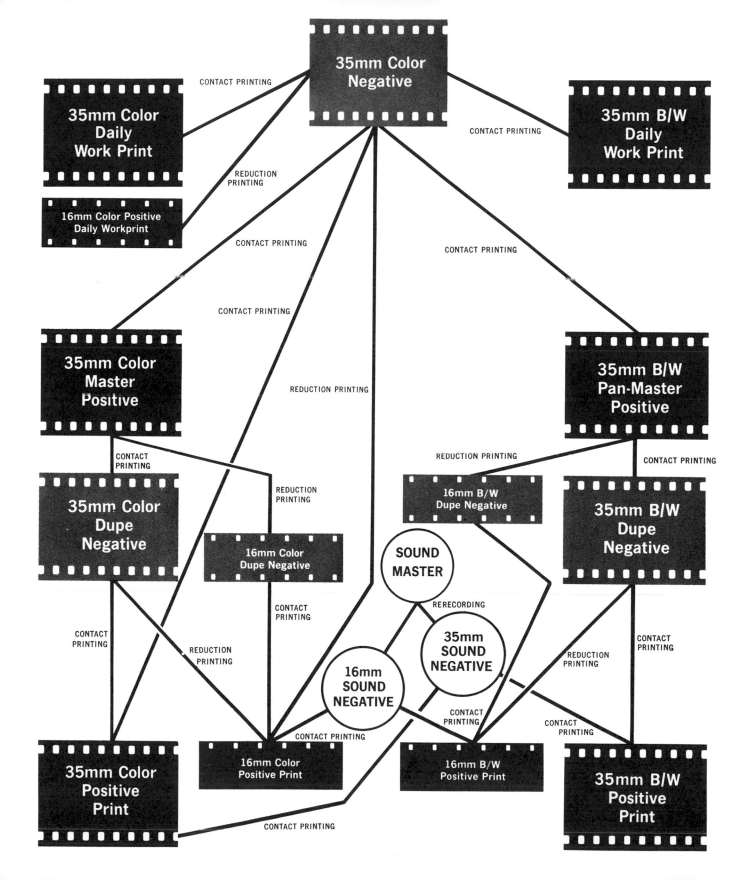

EDITING

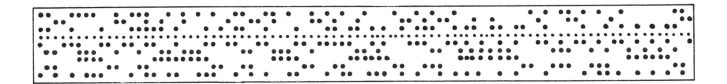

The film editor, the unsung hero of the motion-picture industry, assumes a major part of the responsibility for the completed film presentation. More often than not the editor's judgment and ability with the splicer are the determining factors that separate the hit from the bomb.

Assembling, sorting, and splicing the film exposed for even a 1-minute television commercial is in itself a Herculean task. It is not unusual for a film editor to handle, sort, and discard approximately ten to twenty times the amount of film that is actually used in the final production. The greater amount of the unused footage consists of **outtakes** and the miscellaneous mattes, wipes, titles, and other film elements required for optical-combination purposes. Approximately 90% of the film needed to produce the composite print is never seen by the viewer.

The **sound reader**, for analyzing sound tracks; the **synchronizer**, for measuring frames of film; the **splicer**, for joining separate lengths of film; the scissors, for cutting them apart; and the moviola, for synchronizing the picture with the corresponding sound track, are only some of the tools used by the film editor. Original negatives, dupes, printing material, and fine-grain film are handled with white cotton gloves. Other working tools, on or within arm's reach of the editor's underlit table, are the white or red wax pencils, used for marking work prints; conversion charts, indicating frames and footages in terms of projection time; and wipe charts, from which the editor can select any one of a wide variety of wipe patterns for transitional effects.

Between the cataloging and screening of the first dailies, or rushes, and the distribution of the release prints the editor either prepares or is responsible for the production of the rough cut, work print, optical print, composite print, and, finally, release print. The **rough cut** consists of all the good takes selected during the initial screening. Included are both live-action and animation sequences, which are spliced together in their proper order but not necessarily to the exact length. The **work print** is used during production. In this print the picture information on one reel and the sound track on another reel are synchronized and projected simultaneously in what is referred to as an **interlock**. The **optical print** is the first combination of all the scenes in the production along with visual and/or special effects. This print is produced on the optical printer and used to check the visual continuity. The **composite print** contains both the picture information and the sound track on the same length of film. The **release print** is the final, color-corrected print that is distributed to theaters and television stations.

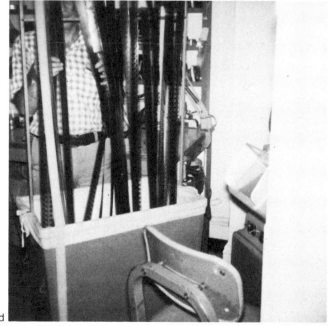

5-1. (a) The synchronizer and sound reader. (b) Analyzing the sound track with the synchronizer and sound reader. (c) The editor and the moviola. (d) The trim bin.

FILM CONVERSION TABLE

TIME	FRAMES	8mm Ft.	16mm Ft.	35mm Ft.	WORDS
1s.	24	.3	.6	1.5	2
2s.	48	.6	1.2	3.0	4
3s.	72	.9	1.8	4.5	7
4s.	96	1.2	2.4	6.0	9
5s.	120	1.5	3.0	7.5	11
6s.	144	1.8	3.6	9.0	13
7s.	168	2.1	4.2	10.5	16
8s.	192	2.4	4.8	12.0	18
9s.	216	2.7	5.4	13.5	20
10s.	240	3.0	6.0	15.0	22
11s.	264	3.3	6.6	16.5	24
12s.	288	3.6	7.2	18.0	26
13s.	312	3.9	7.8	19.5	28
14s.	336	4.2	8.4	21.0	30
15s.	360	4.5	9.0	22.5	33
16s.	384	4.8	9.6	24.0	35
17s.	408	5.1	10.2	25.5	37
18s.	432	5.4	10.8	27.0	39
19s.	456	5.7	11.4	28.5	41
20s.	480	6.0	12.0	30.0	44
21s.	504	6.3	12.6	31.5	46
22s.	528	6.6	13.2	33.0	48
23s.	552	6.9	13.8	34.5	50
24s.	576	7.2	14.4	36.0	52
25s.	600	7.5	15.0	37.5	54
26s.	624	7.8	15.6	39.0	56
27s.	648	8.1	16.2	40.5	59
28s.	672	8.4	16.8	42.0	61
29s.	696	8.7	17.4	43.5	63
30s.	720	9.0	18.0	45.0	65
31s.	744	9.3	18.6	46.5	67
32s.	768	9.6	19.2	48.0	70
33s.	792	9.9	19.8	49.5	72
34s.	816	10.2	20.4	51.0	74
35s.	840	10.5	21.0	52.5	76
36s.	864	10.8	21.6	54.0	78
37s.	888	11.1	22.2	55.5	80
38s.	912	11.4	22.8	57.0	82
39s.	936	11.7	23.4	58.5	85
40s.	960	12.0	24.0	60.0	87
41s.	984	12.3	24.6	61.5	89
42s.	1008	12.6	25.2	63.0	92
43s.	1032	12.9	25.8	64.5	94
44s.	1056	13.2	26.4	66.0	96
45s.	1080	13.5	27.0	67.5	98
46s.	1104	13.8	27.6	69.0	100
47s.	1128	14.1	28.2	70.5	102
48s.	1152	14.4	28.8	72.0	104
49s.	1176	14.7	29.4	73.5	106
50s.	1200	15.0	30.0	75.0	108
51s.	1224	15.3	30.6	76.5	110
52s.	1248	15.6	31.2	78.0	112
53s.	1272	15.9	31.8	79.5	115
54s.	1296	16.2	32.4	81.0	117
55s.	1320	16.5	33.0	82.5	119
56s.	1344	16.8	33.6	84.0	122
57s.	1368	17.1	34.2	85.5	124
58s.	1392	17.4	34.8	87.0	126
59s.	1416	17.7	35.4	88.5	128
60s.	1440	18.0	36.0	90.0	130
2m.	2880	36.0	72.0	180.0	260
3m.	4320	54.0	108.0	270.0	390
4m.	5760	72.0	144.0	360.0	520
5m.	7200	90.0	180.0	450.0	650
10m.	14400	180.0	360.0	900.0	1300
20m.	28800	360.0	720.0	1800.0	2600
30m.	43200	540.0	1080.0	2700.0	3900
60m.	86400	1080.0	2160.0	5400.0	7800

5-2. Film-conversion table.

In the absence of specific editing rules, patterns, or procedures the editor is expected to be creative and imaginative. His approach to a production varies depending upon the director's interpretation of the script. Since the abilities of the editor and the director are judged collectively, the viewer is assured that the film's subject matter will be presented in the most interesting manner possible. The same script in the hands of different directors would reflect individual tastes and produce completely different flows of picture information; similarly, the same picture information in the hands of different editors would continue to reflect individual tastes and treatments. One editing effort might keep the audience on the edges of their collective seats from opening title to closing fadeout, while a second film treatment might keep the popcorn venders unusually busy.

In addition to synchronizing the picture information with the recorded dialogue, the editor must also plan musical backgrounds that fit the scene's requirements and sound effects that accent specific actions. These sound elements must be synchronized with the corresponding picture and labeled for quick and easy identification. **Leader strips**, spliced to each of the small rolls that carry the sound information, make it easy for the editor to locate specific **sync points**, or reference frames, which are also used by the sound-recording studio to combine all of the sound tracks (dialogue, music, and effects) during the **mixing** process into the composite track.

The imaginative film editor will turn to the carefully cataloged stock shots, if necessary, for lengths of film used in earlier productions. If the needed footage cannot be located in the studio library, commercial stock-shot organizations make available, for a reasonable fee, film footage covering every conceivable subject. **Scratch prints** of the selected film contain deliberate markings to prevent unauthorized duplication. Usable footage, reprinted from the library's negative, does not contain these markings. Similarly, music and effects libraries can provide excerpts from the classics or a rock concert, as well as the pitter-patter of a spring rain or a howling hurricane.

Beginning with the receipt of the first film cans

filled with dailies and up to the first critical review of the release print, the editor must remain objective. He is completely responsible for all postproduction processes, including assembly of picture information, synchronization of sound and picture, optical combination of picture elements, and mixing of sound elements, and his impartial opinions and judgments are usually correct. If he were influenced by either the director or the cameraman, not a single frame of film would ever be discarded. The additional frames of film exposed at the beginning and end of each scene, insurance takes, closeups, and cover shots using different camera angles provide the editor with a great deal of latitude and flexibility. The excess footage is not intended to lengthen the film—it is completely expendable. Editing, often referred to as **cutting**, should be just that—the elimination of unneeded pictures and the tightening of the balance to create a fast-paced, interesting, smooth flow of visual continuity.

Each of the 1,000′ film cans that are stacked on shelves within arm's length of the editor and his assistants looks alike. Similarly, the reels or cores of film within the containers have no distinguishing characteristics that can be used to identify the contents. The editor's sanity is preserved and hours of needless search are eliminated by coding systems that include the use of **edge numbers** imprinted along each length of film. They are duplicated by the laboratory on work prints and other film elements.

Original film must be handled carefully—and only while wearing the white-cotton gloves mentioned earlier. Under no circumstances should original material be projected. The ever-present danger of scratches and abrasions, which would certainly be reproduced in subsequent prints, makes projection not only inadvisable but foolhardy. Torn or pulled sprocket holes would also affect registration during printing.

In assembling scenes and preparing the work print the moviola is used for viewing lengths of film and for synchronizing the sound track with the corresponding pictures. Each scene that is cut into the work print should be evaluated in terms of the total effect, so newly inserted scenes should be projected along with one or more scenes that precede or follow the added footage. Details that may be unnoticed on the moviola's small viewing surface immediately become apparent on the large projection screens. If necessary, the editor can lengthen, shorten, or replace an entire scene or an unusable portion at this point with provisional footage (cover shots and insurance takes, for example).

As an economy measure the editor often works with black-and-white material rather than with color prints. The savings are significant, and the disadvantages—having to rely on judgment for scene-to-scene color balance, are negligible.

Screening completed sequences of a film while it is in the work-print stage is important for a number of reasons. Audience reaction—frank comments and fresh viewpoints—may require a reappraisal of earlier efforts. Because of the editor's familiarity with the subject matter the need for new approaches and/or changes in timing often goes completely unnoticed.

For the editor engaged in the production of commercial films for television, analyzing the sound track for the timing of animated actions and lip sync presents additional problems. More often than not animation sound tracks are recorded before the animator sharpens his pencil. These prerecorded sound tracks, guided through a sound-reproducing device known as a **reader**, provide the editor with frame and footage counts indicating the amount of time allotted for a particular action as well as for the recorded dialogue. Each sentence is broken up into words, and the phonetic spelling is entered in the clear area of the magnetic film that carries the sound track. Each word, in turn, is reduced to syllables whose length is measured by means of the synchronizer. Transferred to **bar**, or **lead**, sheets by the animation director, the corresponding mouth-action positions drawn by the animator will be amazingly realistic because of these lip-sync frame counts.

Unlike the presync procedure, in which the animator works from an analyzed sound track, the dialogue, music, and sound-effects tracks for the postsync production are recorded after the animation has been completed and photographed. In this case the editor must match the film to the sound track rather than the sound track to the corresponding picture. Unlike the almost mechanical synchronization of picture and sound in the presync production, postsync processes place an added burden on the editor. Compromises in synchronization are to be expected since the editor must do the best he can with the available material. The more precise presync technique is preferable in practically all instances to the postsync animation-recording process. The sound track that establishes moods for the corresponding pictures in a feature production is especially important for an animated cartoon. In a world of fantasy populated by make-believe characters sound effects and musical backgrounds accent the animated antics and add touches of realism.

Editing procedures for 16mm films are similar to those for the wider 35mm-film presentations. The process of assembling the picture information and synchronizing the music and effects with the corresponding images is the same for both film widths. The main differences occur after the work print has been completed. Unlike the optical combination of the printing elements in the 35mm production, the 16mm work print is divided into two rolls of approximately equal size, identified as **A & B rolls**. The odd-numbered scenes are spliced together on one roll, and the even-numbered scenes on the other. Each scene on both rolls is separated from the next by a length of black leader. The amount of leader spliced between scenes 1 and 3 in the A roll is equal to the number of frames in scene 2 on the B roll. The same procedure is followed in preparing the B roll: the length of leader between scenes 2 and 4 is equal to the length of scene 3 on the A roll. This checkerboard pattern renders splices between scenes invisible during laboratory printing processes, in which each scene is duplicated alternately on a single roll. Transitional effects such as cross-dissolves and fades and the superimposition of titles can also be achieved with A & B printing

processes. The composite print will have each scene from the separate rolls in its proper order, color-corrected, and complete with respect to transitional effects and title inserts.

Many service organizations are equipped to handle opticals in every film width from 35mm to 16mm, Super 16mm, and Super 8mm. The difference in cost between optical printing and laboratory A-&-B-roll processing for the narrower-gauge films is usually the determining factor in selecting printing processes.

The completion of a work print for an animated-cartoon or feature-film production and the subsequent optical combination of picture elements do not signal the end of the editorial effort. The composite print, in which picture and sound appear on the same length of film, is still subject to final approval before the release prints are ordered in quantity and distributed to theaters and exhibitors. With final approval the accumulation of film cans filled with images and sounds are carefully packed and stored in temperature-controlled film vaults. If the occasion arises, the film editor may ask an assistant to recover a specific scene, musical background, or sound effect for use in another production.

Assembling individual scenes into a smooth flow of visual continuity, synchronized with a musical background and sound effects, is the editor's responsibility. The production of a well-integrated composite is anything but a routine mechanical procedure. At every stage of production from cataloging dailies through the mix and optical processes the editor must exercise judgment and be imaginative, patient, and diplomatic. Ever mindful of the old adage, "A soft answer turneth away wrath," the film editor must sometimes be a better actor than the people who appear in the film that he is cutting. The editor should have all of the qualities that the State Department expects in a diplomat—a sense of humor is a prerequisite!

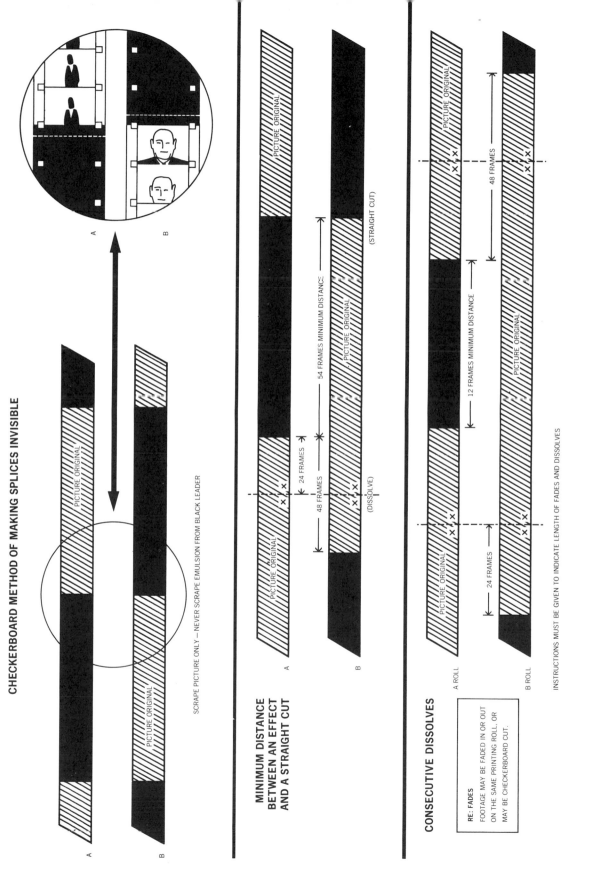

CHECKERBOARD METHOD OF MAKING SPLICES INVISIBLE

PICTURE ORIGINAL

PICTURE ORIGINAL

A

B

SCRAPE PICTURE ONLY — NEVER SCRAPE EMULSION FROM BLACK LEADER

MINIMUM DISTANCE BETWEEN AN EFFECT AND A STRAIGHT CUT

PICTURE ORIGINAL

PICTURE ORIGINAL

A

B

24 FRAMES

48 FRAMES

54 FRAMES MINIMUM DISTANCE

(DISSOLVE)

(STRAIGHT CUT)

CONSECUTIVE DISSOLVES

PICTURE ORIGINAL

PICTURE ORIGINAL

A ROLL

B ROLL

24 FRAMES

12 FRAMES MINIMUM DISTANCE

48 FRAMES

RE: FADES

FOOTAGE MAY BE FADED IN OR OUT ON THE SAME PRINTING ROLL, OR MAY BE CHECKERBOARD CUT.

INSTRUCTIONS MUST BE GIVEN TO INDICATE LENGTH OF FADES AND DISSOLVES

5-3. Preparation of 16mm original reversal A & B rolls. (Courtesy of General Film Laboratories.)

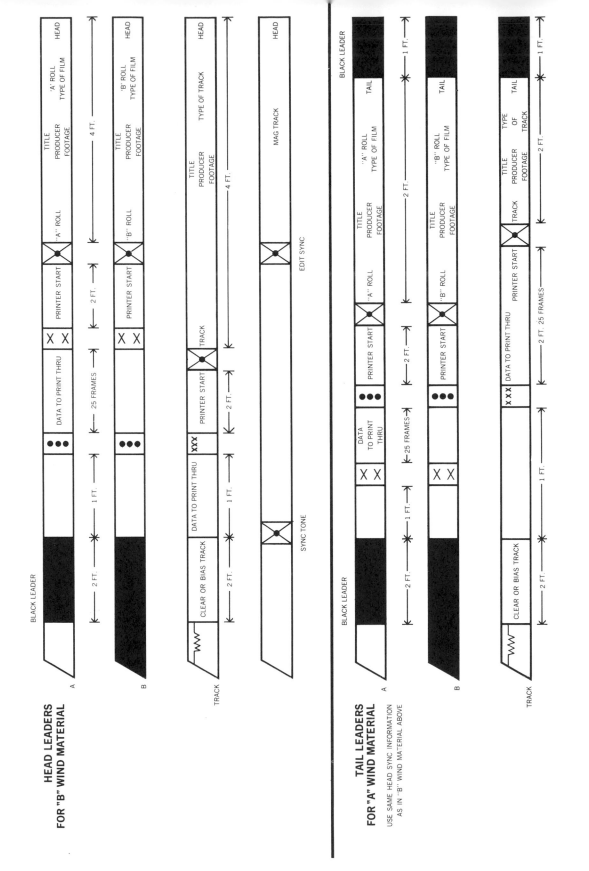

5-4. Preparation of 16mm printing leaders. (Courtesy of General Film Laboratories.)

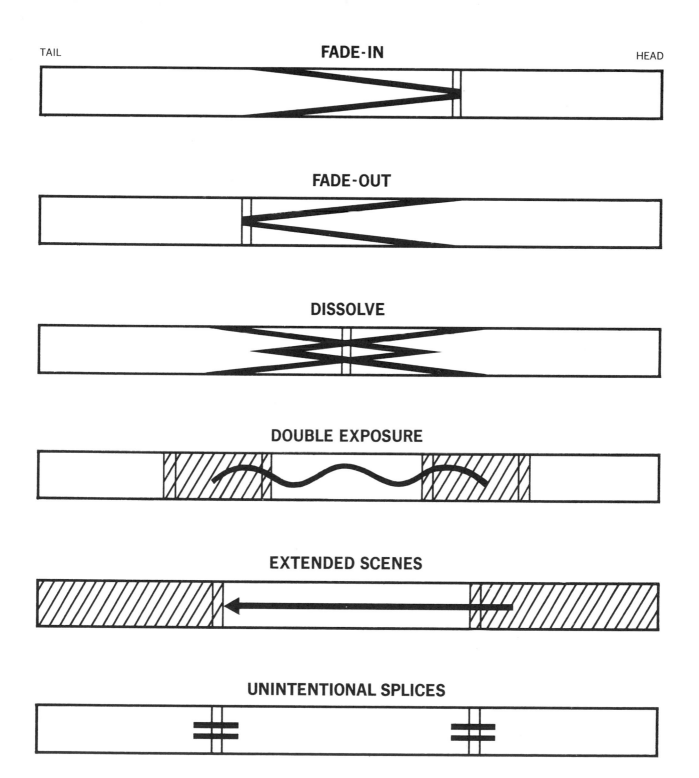

5-5. Work-print markings for special effects. (Courtesy of General Film Laboratories.)

EDITING GLOSSARY

A & B printing A method of combining picture information from two separate rolls of 16mm film. Arranged in checkerboard fashion, all of the odd-numbered scenes are cut together to form the A roll, and the even-numbered scenes are similarly spliced into another B roll. Each scene is duplicated sequentially during laboratory processing. Transitional effects such as cross-dissolves and fades are added during the printing process aiong with superimpositions to produce the composite print.

A & B rolls The rolls into which the scenes are separated in editing 16mm film material. The odd-numbered scenes are spliced together into the A roll, and the even-numbered scenes into the B roll. Lengths of opaque black leader are spliced between each scene. The length of leader between scenes 1 and 3 on the A roll, for example, is equal to the length of scene 2 on the B roll; similarly, the length of leader between scenes 2 and 4 on the B roll is equal to the number of frames in scene 3 on the A roll. During the laboratory-printing process the scenes from each roll are printed sequentially to produce the composite print. Effects and superimpositions, such as titles and other graphic material, are spliced into a C roll and added to the specified frames.

advance, sound track to picture The number of frames between any specified point in the sound track and the corresponding picture. In a composite 35mm motion-picture print the sound track precedes the corresponding picture by 20 frames; in a 16mm print, by 26 frames. The position of the gate in the motion-picture projector, through which the picture passes while the sound-track area is scanned by the projector's exciter lamp, places the picture 20 frames ahead of the corresponding sound. Advancing the sound track in relation to the picture places each element in its proper projection position simultaneously.

anamorphic print See **anamorphic print** in the optical-effects glossary.

answer print See **optical print** in the optical-effects glossary.

background, sound Music and/or sound effects synchronized with the picture information. The background music sets the mood for the corresponding picture. The sound effects complement the visual information and accent actions needing sound support.

bar sheet, or **lead sheet** See bar sheet in the animation glossary.

beat (1) The tempo of the background music. (2) The pacing for an animated action. A visual effect can be emphasized by accenting the beat.

bloop A small, triangular-shaped piece of tape or plastic that is placed over the sound-track area at splice points. The mask covers the objectionable sound that occurs during projection whenever a splice is encountered.

blowup See **blowup** in the optical-effects glossary.

bridge A transitional scene or musical effect designed to improve the visual and/or sound continuity of a production.

butt splicer A type of splicer designed to eliminate the possibility of out-of-focus frames at a splice point or of frame loss, which would throw a double-system sound track out of sync.

checkerboard editing The preparation of A and B rolls for processing. See **A & B printing** and **A & B rolls**.

click track A cue track used by the film editor in the absence of the sound track that will subsequently be synchronized with the corresponding picture. It is recorded with little regard for quality and used to establish a beat or to provide the timing for an action sequence.

composite master positive A fine-grain composite print, either black-and-white or color, that is used for making composite dupe negatives.

composite print A length of film combining picture and sound elements.

composite reversal original An original positive print, produced by reversal processing, that contains both picture and sound track.

cording See **cording** in the optical-effects glossary.

cover shot, or **insurance take** The reshooting of a scene to change the performance, camera angle, or set lighting.

cue sheet The frame count provided by the film editor, which serves as a guide for the optical combination of picture elements.

cut The abrupt end of a scene or action. The following scene is spliced at the cut point.

cutaway A scene inserted into a sequence in order to add suspense to the flow of action. It might include a closeup of the performer's eyes or hands or the reaction of another performer in the scene.

cutback The scene that follows an insert shot. It represents a continuation of the story line and a return to script continuity.

dailies, or **rushes** The processed results of film exposed during the previous day's photographic efforts.

edge numbers Identifying numbers imprinted at 1' intervals along the edges of the film. They are duplicated in the print material and used by the optical-layout man in the production of the optical print and by the film editor in matching the negative to the work print.

Edicomp See **Edicomp** in the computer glossary.

Editec See **Editec** in the computer glossary.

editing The sorting, assembly, splicing, synchronizing, and general preparation of the various film elements into a smooth flow of picture and sound information.

editor The person responsible for assembling individual scenes, arranging them in the order indicated in the script, and synchronizing the sound elements (dialogue, music, and effects) to the corresponding picture. In addition to supervising the preparation of the work print the editor is also responsible for the optical, special, and transitional effects produced by optical processes.

editorial synchronism The synchronization of sound and picture elements during the editorial process. The separate sound track is aligned at all points with the corresponding picture while the work print is assembled. Compare **projection synchronism**.

feed reel The plastic or metal reel on which the original film or tape is wound.

fine cut The assembly of a number of scenes in proper order, with each scene cut to its final length.

fine-grain See **fine-grain** in the optical-effects glossary.

flashback An inserted scene that takes place at an earlier time.

flash frame See **flash frame** in the optical-effects glossary.

foil Pieces of aluminum foil applied to the film at specific points by the editor. They are used to locate cut points when the picture information is transferred to videotape. The pieces of foil activate a pickup device that places an electronic cue on the videotape.

grading See **grading** in the optical-effects glossary.

guide track A temporary sound track used for synchronization purposes in the absence of the actual recording that will subsequently be combined with the film's picture information.

head out The position of the leader strips in relation to the film wound on a reel. The leaders are wound on the outside of the film roll ready for threading in a moviola or projector.

high-contrast film See **high-contrast film** in the optical-effects glossary.

hot splicer A film splicer with a thermostatically controlled heater that maintains a constant 100° temperature. Fitted with tungsten-carbide inserts for cutting, the splicer also has tempered-steel springs that position the film precisely during the scraping and splicing process.

insert scene A short length of film—a closeup of a performer's hands or eyes, for example—that is intercut between two related scenes in order to help the visual continuity.

insurance take See **cover shot**.

intercutting The alternation of scenes from two related sequences to give the impression of simultaneity.

interlock See **interlock** in the computer glossary.

internegative See **internegative** in the optical-effects glossary.

interpositive See **interpositive** in the optical-effects glossary.

invisible splice A type of splice used for preparing A and B rolls. The film is positioned in the splicer, and the first cut is made on the frame line. The second cut, on the film to be joined, is made in the picture area. Since scrape markings would be reproduced, the film must be handled carefully. The splice is lined up with the leader strip and does not show in the processed film.

jump cut The shortening of a scene by eliminating a number of expendable frames without affecting the continuity of the action.

leaders Lengths of film that are spliced at the beginning and end of a roll of film. The head leader is threaded through the projector's transport mechanism, eliminating the need to handle any part of the picture itself. Numbers are marked in reverse order on these leader strips at 1′ intervals to assist the projectionist.

lead sheet See **bar sheet** in the animation glossary.

library footage, or **stock shot** An entire scene or length of film from an earlier production that is used as an insert. Commercial organizations specializing in library footage issue catalogs describing unusual locations and special-effects footage that are available on a rental basis. Sound and music libraries offer similar services. See also **scratch print**.

lip sync (1) A sequence of mouth-action drawings prepared by the animator to fit the film editor's analysis of the prerecorded dialogue. (2) In live action, the synchronization of the sound track with corresponding pictures in which dialogue between performers takes place.

magoptical print A print containing both magnetic and optical sound tracks.

magoptical track A length of film containing both magnetic and optical sound tracks. This type of track is arranged alternately and used for editing purposes.

M & E track A sound track containing music and sound effects only. The accompanying dialogue is recorded separately. Separate tracks are especially useful if the dialogue is to be dubbed in a foreign language.

marriage The combination of picture and sound-track information in the composite print.

master positive See **master positive** in the optical-effects glossary.

matching negative The delicate process of cutting and matching the original picture negative to the edited work print.

mix The rerecording of all sound elements (dialogue, music, and effects) into a single sound track that is used to prepare the composite print.

modulation The sound patterns in variable-area and variable-density sound tracks.

mood music A musical background that helps to establish a mood or setting for the corresponding picture.

moviola A viewing device used by the film editor to synchronize the sound track and the corresponding picture.

mylar-tape splicer, or **guillotine splicer** A splicer in which mylar tape is used to join lengths of film in either butt, overlap, or diagonal splices without losing frames. It is used only for editing purposes.

notch A cutout area along the edge of the film, used by optical technicians and laboratory timers to indicate changes in printing lights.

optical effect See **optical effect** in the optical-effects glossary.

optical print, or **answer print** See **optical print** in the optical-effects glossary.

optical recording The optical sound patterns in the sound-track area of the film. During the projection process the modulations are electrically converted from light impulses into audible sound. See also **variable-area** and **variable-density sound tracks**.

original picture negative The negative produced by the laboratory from the in-camera raw stock used during the actual filming process.

outtake A duplicate scene not used in the final film.

panchromatic master positive, or **pan master** A black-and-white print produced from a color negative on panchromatic film and used to make black-and-white duplicate negatives.

picture duplicate negative, or **dupe negative** A negative produced from positive materials such as black-and-white, color, or separation master positives.

postsynchronization The recording and synchronization of a sound track with the picture after the filming process has been completed. The film editor must provide arbitrary timings for animated actions and lip-sync sequences. Compare **presynchronization** and **reading** (in the animation glossary).

presynchronization The recording of a sound track before the filming process takes place. In animation the recording of the sound track usually precedes any of the actual production processes. The film editor provides the animator with frame counts, derived from the sound-track analysis, which show the number of frames allotted for an action and the frames in which dialogue takes place. Compare **postsynchronization** and **reading** (in the animation glossary).

print A positive picture produced from an original or dupe negative. The colors and tonal values of the original subject matter are reproduced in the processed print.

processing The production of visible images, by means of laboratory printing processes, on lengths of previously exposed film. Normal development of a length of black-and-white motion-picture film yields a negative image. The tonal values in the developed film are exactly opposite to those in the original subject matter. With proportionately greater silver deposits in the lighter areas black areas become clear in the negative-film frame, and the lighter areas appear proportionately denser. In the processed print the tonal values are similar to those found in the subject matter at the time of exposure. By comparison, in-camera reversal stock yields a direct positive. To produce the positive image, the film is developed in two stages. In the first stage the developed silver in the emulsion is exposed and bleached away by chemical action. In the second step the remaining unexposed silver is reexposed, redeveloped, and fixed to yield the positive picture information. A chemical reaction referred to as **coupler development** is the basis for the reproduction of color values. The exposed silver halide in the color film forms metallic silver during the first stage of the developing process. The oxidizing developing agent combines with the chemical coupler to produce insoluble dyes, which are proportional to the amounts of metallic silver formed during the first stage of the developing process and which reproduce the colors in the original subject matter. The color values of the negative images are complementary to those in the original. When the coupler components are extracted from the developing solutions, the positive images produced by color-reversal processing are obtained by developing each of the three emulsion layers separately. If the couplers are added to the emulsion layers of the original in-camera, color-reversal film by the manufacturer, however, the film is developed in a single solution to produce the direct-positive images.

projection synchronism　The alignment of sound and picture information on the same length of film. The sound precedes the corresponding picture by 20 frames in a 35mm print and by 26 frames in a 16mm print. Compare **editorial synchronism**.

quick cut　A moving montage effect produced by splicing together a number of short lengths of film. In most cases, despite the related subject matter, there is no attempt to establish visual continuity.

reader, sound　A device for reproducing sound that is used for editing and analyzing sound tracks. It has a revolving drum over which the track area passes to make contact with the reproducing head.

reduction print　See **reduction print** in the optical-effects glossary.

reel　A plastic or metal spool used for winding film.

release negative　A negative containing both sound and picture information, from which the release print is produced.

release print　The final version of a film, complete with respect to sound and picture information, that is distributed to theaters for exhibition.

rerecording　See **mix**.

reversal original　See **reversal original** in the optical-effects glossary.

reversal processing　See **reversal processing** in the optical-effects glossary.

rewind　A device attached to the film editor's table that is capable of supporting a number of feed and take-up reels.

rough cut　A temporary arrangement of scenes, cut to their approximate length, used to check the visual continuity.

rushes　See **dailies**.

scene　An action or incident, filmed in an appropriate location or setting, that forms part of a related sequence in a film presentation.

scraper　The device on the splicer that is used to remove the emulsion from the part of the film frame in which the splice is to be made. Film cement is applied to this area, and the length of film that is to be joined is pressed into position to complete the splice.

scratch print　Library footage or stock shots made available by commercial organizations. Stock footage of every conceivable subject can be leased by the editor. The film is deliberately scratched to prevent unauthorized duplication. Unscratched footage is produced from the original negative. Some libraries specialize in music and sound effects.

scratch track A temporary recording prepared for synchronization purposes only. The film editor works with this type of sound track if the original recording is not readily available.

segue A musical transitional effect.

separation positive See **separation positive** in the optical-effects glossary.

sequence A grouping of related scenes that are spliced in proper order and complete with respect to visual continuity.

shot A specific scene or the action occurring within the scene.

skip frame See **skip frame** in the optical-effects glossary.

sound effect Any sound that helps lend credibility to the corresponding picture.

sound reader See **sound reader** in the animation glossary.

sound track The recorded sound that complements the corresponding picture information. The composite sound track contains dialogue, musical backgrounds, and sound effects. In 35mm film the sound track is positioned longitudinally in a narrow area on the left side of the film; in 16mm film it is reproduced along the right side of the frame. See **advance, sound track to picture**.

sound-track advance See **advance, sound track to picture**.

sound work print A sound track containing sections of dialogue, music, and effects. The intercut track is synchronized with the corresponding picture information in the work print.

splicer An editing accessory used to join lengths of film.

split reel A spool used for winding film. One section of the reel unscrews from the other, enabling a roll of film to be removed from the reel without rewinding.

start mark, or **sync mark** An X drawn with grease pencil across specific frames in both the sound track and the corresponding picture to indicate a common synchronization point.

stock shot See **library footage**.

sync mark See **start mark**.

synchronism The alignment of the sound track with the corresponding picture, which is on a separate reel. See **editorial synchronism** and **projection synchronism**.

synchronization The precise relationship between the sound track and the corresponding picture information.

synchronizer A sprocketed device used by the editor to measure lengths of film.

take See **take** in the animation glossary.

take-up reel The reel on which the film is wound after it has been run through a synchronizer, moviola, or projector shuttle.

tightwinder A guide roller and hinged metal spring-tension arm used for winding film quickly on cores or hubs rather than on reels or flanges.

transfer The rerecording or duplication of a magnetic sound track on sprocketed magnetic sound film.

transitional effect See **transitional effect** in the optical-effects glossary.

trim An unused portion of a scene.

variable area sound track   An optical sound track. The modulations, corresponding to the variations of audible sound, form irregular patterns in the sound-track area. Compare **variable-density sound track**.

variable-density sound track   An optical sound track in which the modulations form constantly changing light gradations ranging in density from intermediate shades of gray through black. Compare **variable-area sound track**.

wide-screen print   A print in which the picture information within the film frame is wider than the standard 1.33:1 aspect ratio. Unlike an anamorphic print, in which the image is squeezed and returned to normal proportions during projection by means of a complementary deanamorphic lens, a wide-screen print need not contain a squeezed image.

wild track   A temporary sound track, containing dialogue and/or music, that approximates the contents of the recorded sound track that is synchronized with the corresponding picture. This time-saver is used by the editor for postsync productions or as a substitute for the actual track if it is not available.

work print   The assembly and orderly arrangement of scenes, cut to length and marked with a grease pencil to indicate the location of special effects, transitions, and superimpositions. It is synchronized with the accompanying sound track and used as a guide for the production of the optical, or answer, print.

zero cut   A method of hiding splices in A-&-B-roll editing. Each scene is extended by two frames at the beginning and end, and black leader is spliced between the scenes. Cuts between scenes are smooth because the overlapping footage eliminates splices. See **A & B rolls** and **A & B printing**.

PART 2:

COMPUTER-CONTROLLED TECHNIQUES

INTRODUCTION

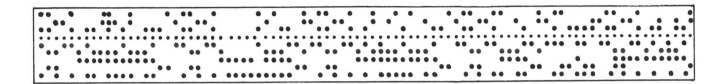

A computer operating in conjunction with a graphic-output device can make even a Rembrandt or a da Vinci throw down his brushes and seek new outlets for his talents. The computer can draw a straighter line and a truer circle. It can create as well as copy. It can be used to produce simple line drawings or images of great complexity. It can generate graphics that could not possibly be rendered or duplicated by hand. If the mind's eye could visualize the application of hundreds of thousands of brushstrokes on canvas simultaneously, it could relate to the mathematical potential of the computer—the most powerful tool ever devised for the production and display of graphic material. The logical question seems to be, "Is the artist the creator or is it the computer programmer?" The artist is, of course, creative, imaginative, reasonably intelligent, and motivated. The same adjectives can usually be used to describe the programmer, who is also logical, precise, and methodical. Some compromise is necessary. A distinction can be made between the person who initiates the concept—the artist—and the person who pushes the button or turns the dial that transforms the concept into a reality—the programmer.

Turning a concept into a reality can be accomplished in a number of ways. The picture generated by a computer may consist of hundreds of thousands of values or symbols, each of which is as-signed an electronic position or numerical equivalent. These image-forming elements may either be stored in the computer's **memory bank** and displayed at some future date or reproduced instantaneously and displayed on the face of the **cathode-ray tube** (CRT) in a manner similar to standard television transmission. The image in a standard television receiver is formed over 525 **scanning** lines, each of which contains approximately 450 dots, or picture elements. Each dot can be exposed, like film in a camera, by an encoding device that assigns all of the component parts of the picture (the input signals) to a specific position in the display. Each dot also has a light-intensity potential ranging from black to white in incremental steps. With 0 representing black, and 10 white all of the intermediate numbers can be translated into corresponding shades of gray. With the television-scanning system each of the scanning lines on which the image is formed is traced by an electron beam that sweeps across the tube face 30 times each second. The subject matter viewed by the TV camera is constantly being retraced, or refreshed.

Another method for generating and displaying images electronically is referred to as the **planar coordinate process** (PCP). The PCP image is represented by numerical equivalents, with each number assigned a precise position on the display grid. The component parts of the generated image con-

6-1. *Digital Mona Lisa* by H. Phillip Peterson. It was produced as a digital plot by a scanner and two Control Data Corporation computers. Each small unit contains two decimal digits that report the density at various points on a color-projection-slide copy of the painting. The painting is composed of square cells measuring 0.115″ on a side; it is 256 cells wide and 390 cells high, with about 100,000 cells in all. Cells are marked off by ticks in the horizontal lines between rows. Each cell contains two decimal digits, the magnitude of which is proportional to the density at the corresponding point on the slide as measured by a CDC 160 computer driving a special high-resolution jumping-spot scanner. The number in each cell is the result of scanning eight points in each tiny area and averaging their densities. The digits are plotted by an incremental plotter driven on-line by a CDC 3200 computer. The font is designed so that the larger the pair of digits, the darker they appear to the human eye. Up close the viewer sees what the computer "sees"—a number field—but at a distance of 30′ the shading is comparable to that of a newspaper photograph. With 100 gray levels the computer-generated display has better definition than the average newspaper reproduction. About 61 plotter steps are needed to make each digit. The completed picture required approximately 16 hours of plotting time and consists of 16,000,000 plotter steps (300 per second). It was completed in one continuous computer run. (Courtesy of Control Data Corporation.)

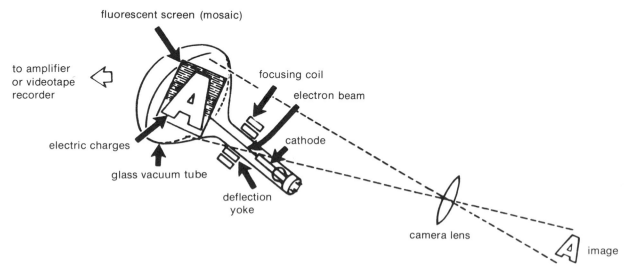

6-2. Television pickup tube.

sist of a series of short, straight lines. Even a circle is formed by line segments that are arranged to produce the geometrical figure. Each of the intersecting points in the grid is represented by both x- and y-coordinates—the first indicating the horizontal position, and the second the vertical position. Even the shortest line possible in this type of arrangement must be represented by four numbers—two coordinates for the beginning point of the line and two for the end of the line. Controlled by an electronic computer, the **microfilm recorder** can plot points and, following the instructions fed by the programmer, draw lines a million times faster than any person. The recorder is basically a camera and a display tube that understands only simple instructions. It can display a spot of light at specific coordinates or create a series of lines in different parts of the grid.

A repertoire of preprogrammed instructions—the programming **language**—can be custom-designed around industry terminology. Computational routines and subroutines, operating in parameters, can be arranged hierarchically so that complicated operations are greatly simplified. This sophisticated language capability makes possible the formation and display of basic image-forming elements at incredible speeds.

The development of computer-generated animation techniques began in the early 50s and continued through the 60s. Computer-generated sequences of motion projected on large studio screens magnified the effectiveness of the automated processes, pointed out the potential, and dictated the direction for future experimentation, which is continuing at an ever-accelerating pace. Credit for the development of the technique must be shared by many. Dr. Kenneth Knowlton of the Bell Telephone Laboratories, Lee Harrison III of Computer Image Corp., Al Stahl of Animated Productions, Inc., James and John Whitney, special-effects technicians for feature productions such as *2001: A Space Odyssey*, Steven Rutt and Bill Etra of Rutt Electrophysics, Dr. Philip Mittelman of the Mathematical Applications Group, Inc., Marceli Wein of the National Research Council of Canada, Stephen A. Kallis, Jr., of the Digital Equipment Corp., E. E. Zajac, Stan Vanderbeek—these are only a few of many pioneers. A more complete list would include numerous lesser-known experimenters outside the United States.

The use of computers to generate movement evolved accidentally—the result of early experiments by scientists and engineers who used prototypes of today's sophisticated systems to analyze objects in motion and to create/or simulate conditions with graphics that could be displayed on electronic monitoring devices. The first method for modifying and distorting images can be traced back to the middle 40s—a by-product of radar-related experiments. The technology for controlling the degree of distortion was developed shortly thereafter.

The availability of computer systems and the generally reluctant acceptance of their capabilities by a skeptical industry have forced a comparison and reappraisal of the relative merits of conventional animation-production processes. It has also become necessary to redefine the generic term itself: until now "animation" could be defined as a stop-motion photographic process in which the shape and position of static artwork changes in relation to the camera position on successive frames of film. When the processed film is projected at normal viewing speeds, the changes become evident and the illusion of movement is created. New definitions must include the function and position of the programmer and the computer as well as the animator in the overall production process.

There are two basic electronic techniques for creating the illusion of movement. In the first the animation stand's movements and functions are programmed and controlled by the computer. The motion sequence is programmed to the computer via a teletype. The perforated-paper tapes carrying the programmed instructions are fed to a reader that controls all of the animation stand's movements and functions. The graphics for this animation-recording system usually consists of static artwork with cel overlays. The second technique is used to generate and program continuously variable images, which are displayed on a CRT (cathode-ray tube). The artwork may either be prepared with conventional processes and "exposed" by the critical lens of a television camera, or the graphics can be generated directly on a CRT by converting each of the component image-forming lines into a series of corresponding electronic equivalents. The two basic techniques can be subdivided into four major classifications: **indirect, or off-line, computer-controlled animation recording**; **direct, or on-line, computer-controlled animation recording**; **computer-controlled static images for CRT display** (also referred to as static computer-generated animation imaging, static computer imaging, or static display images); and **computer-controlled continuously variable imaging** (also known as continuously variable display images or dynamic computer-generated animation imaging)

In developing these computer techniques researchers and experimenters considered the artist's needs and the most practical ways of satisfying them. The approach was a wise one, since it allowed even the most rebellious artist to feel as creative and comfortable behind the computer's programming console as he might at his underlit drawing board. By no stretch of the imagination is the animator a technician first and then an artist. Working as a unit, the computer not only challenges the animator's most imaginative efforts but augments them and responds to the artist's slightest whim in less time than it takes to sharpen his pencil. Using familiar tools as well as equally familiar industry terminology, the animator can generate and control movement, shape, and color simultaneously and instantaneously.

Motion sequences that might drive an animator up a wall are taken in stride by the computer. The constantly changing perspective of a solid object revolving in space as the camera zooms from a long-shot position to a closeup that fills the entire frame is the type of problem that is certain to add a few gray hairs to an animator's head. Creating the positive line drawing that is used as the reference input and programming the component parts of the drawing in their precise position in the frame are simple procedures for the computer. Regardless of their complexity, the image-forming lines can be set in motion by merely touching a key, pushing a button, or turning a dial on the control console. The mathematics worked out by the computer causes the lines to move from one position in the frame to the next

position in compliance with the animator's instructions.

How does one issue instructions to or communicate with a computer? In computer language, of course. The computer's vocabulary is limited only by the needs of the animator specifically, and by the industry generally. The computer is programmed to respond to industry terminology—if the animator wants a zoom-and-pan effect from one position to another, he indicates that information by means of a **teletype**. The resulting perforated control tape that carries the information is fed to a **reader** connected to the animation stand's positioning controls. The reader takes over, and the zoom-and-pan effect is produced at the proper point and in the prescribed number of frames. Cinetron and Fortran, described in detail in the section on animation and electronic imaging, typify completely different languages developed for computer programming. The **Cinetron** language programs the operation of equipment and the recording of data; the **Fortran** language, representative of the PCP (planar coordinate process), is used to generate images by assigning corresponding numerical equivalents (x-, y-, z-coordinates) to determinate points in the display grid.

The simplest way of describing a **display** grid in terms of TV raster-scanning lines is to compare it with the open-weave nets used by commercial fishermen. Each knot in the flexible net is comparable to the hundreds of photosensitive picture elements that are spaced evenly along each of the electron tube's 525 scanning lines on which the picture is displayed. The accurate display of an image is controlled by the deflection circuits that dictate the movement of the electron-scanning beam. Modification or distortion of the image is achieved merely by altering the magnetic deflection fields. If the net is stretched along the x-axis, the image-forming elements are distorted horizontally; along the y-axis, vertically. Pulling the net from the center or from one of the corners would completely distort the x- and y-planes as well as with the z-axis, which influences perspective and creates the feeling of depth. The achievement of special effects was not the immediate objective of the pioneers who built the first image-modifying devices now referred to as video synthesizers. Animation, however, seemed to offer a ripe field for commercial exploitation of the new raster-bending systems.

Briefly and nontechnically, the computer systems used for generating images and controlling movement are either **digital** or **analog**. The digital computer is arithmetic and logical: computations are made in sequence on a step-by-step, building-block basis. The input signals used for the formation of an image must be in numerical equivalents that are programmed sequentially. The digital computer consists of a single class of circuit elements that are either on or off, or capable of responding to a programmed command with a simple "yes" or "no" answer. Complex problems are resolved by interconnecting a number of these electronic circuits. The complexity of the setup is proportional to the demands of the programmer. Even the most complex of digital systems, however, can only respond with a simple "yes" or "no" answer—no qualifications. Each phase of a multifaceted problem must be processed in turn. Because of this iterative approach the rate of response is slowed as a system's capacity is increased.

Digital image-generating computers are ideally suited for the complex mathematics needed to produce, program, and display the movement of figures, objects, shapes, and abstract designs. Each of the images required for a sequence of action consists of thousands of points of light. These **bits** are programmed into the computer's memory-storage bank, where they are converted into numerical equivalents. The cathode-ray tube's electron beam, controlled by the programmed instructions fed to the computer, displays the image-forming lines as interconnected points of light. Each of the subsequent moves—from the initial reference input through the final frame of the movement—must be programmed separately and sequentially. The more complex the action, the greater the amount of time needed for programming. Similarly, additional memory units or increased storage capacity is necessary if the planned sequence exceeds the computer's capabilities. The elimination of lines that would normally be invisible behind a solid, dimensional object, for example, would require the use of systems with great computing capacities operating at reduced speeds.

A sequence of action, after being programmed and previewed on the monitor, may still be changed or modified by reprogramming either the entire movement or only the frames that need correction. Once approved, however, the entire movement can be retrieved from the memory units and displayed at real-time rates for reproduction on motion-picture film or videotape. If special effects are needed, the output signal is directed to a video switcher for matteing or for combination with other material to produce a montage or composite effect. Depending upon the system in use, color may be produced directly on the generated display or added optically by means of color-separation prints.

It is quite probable that the next generation of digital computers will have improved storage facilities with provision for immediate access to a large number of preprogrammed routines. The **random-access** storage of graphic information and special effects would be a great budget-saver for producers specializing in variations of Fotomation or limited-

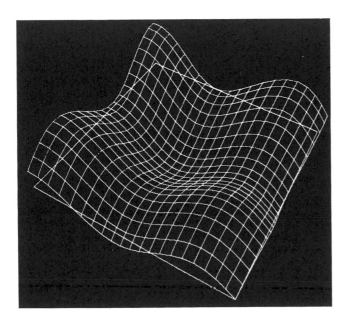

6-3. The deliberate distortion of (x, y, z) coordinates is achieved by altering the magnetic deflection fields.

animation techniques. A Japanese electronics firm has already developed a computer specially designed to create cartoon animation in a limited form. The cartoon character's eyes, mouth, body, and legs are all programmed in a number of positions in the film frame, with each component part designed to move in relation to the others. Because of the numerous variations possible with this type of random programming, predictable results are virtually impossible.

Conversely, asking a computer to program lifelike movement indicates a willingness to compromise for a timesaving, mechanical-looking effect. At the present time the digital computer's "electronic pencil" is not ready for this type of assignment. Analog computer systems, in addition to their mathematical capabilities, are able to differentiate and integrate the continuous input of image-forming elements on the basis of changes in the magnitude of the signals. Unlike the digital computer, in which all the elements are similar, the analog computer contains a grouping of dissimilar elements. Each of these electronic circuits is capable of resolving one of the many individual problems in a complex command simultaneously. Because of the dissimilarity of the circuits the computer's response is not always a positive "yes" or "no"—an occasional "maybe" occurs when that answer seems to be appropriate, and even the "maybe" is subject to change with the introduction and programming of new or additional information.

Analog computers incorporate a number of electronic circuits that are collectively responsible for the formation and movement of some component part of an image. Operating in unison, the circuits or memory-storage units recreate an image in its entirety on the face of the CRT on the basis of the input signals and their conversion to mathematically precise positions in the display. The programmer, very often the animator himself, monitors the output at a rate of 48 or more frames per second. Synchronized with a motion-picture-camera shuttle operating at normal sound speed (24 frames per second), the displayed images can be photographed on film. If the images are to be recorded on videotape, the display rate is increased to 60 frames per second, and the images are recorded at the standard 30-frames-per-second television rate. In this way the images are not only monitored but recorded at real-time rates. **Real time** refers to the actual amount of time needed to perform an action and to record and display that action on an electronic monitoring device. The ability to record in real time represents a great advantage over computer systems that require many minutes merely to form and display a single image in a sequence.

The best features of each of the computer systems—the extreme accuracy of the digital computer and the high-speed functioning of the analog computer—can be combined. The increased capability and efficiency of a **hybrid computer** make it possible to combine existing techniques for producing movement electronically with computer-generated imaging processes to create a number of variations.

There is no real conflict between the animation producer who prefers conventional photochemistry processes and the maverick electronic-image maker. While the hardware for each technique differs considerably, the creative approaches and production techniques are basically similar. Each of the techniques will undoubtedly and inevitably complement and contribute to the improvement of the other. The computer's obvious advantage over conventional production processes lies in its ability to reduce costs by eliminating many of the tedious, time-consuming procedures. It has already demonstrated an ability to meet professional standards in the demanding and critical entertainment and commercial industries. The achievement of full animation is still the primary objective of the designers of electronic image-generating hardware. Prototypes with this capability have already demonstrated the ability to produce dimensional figures directly on color film or videotape. While Disney-type cartoon characters wait in the wings for their entrance cues, the computer-generated, stylized animated figures hog the spotlight and continue their starring roles in commercial television.

Regardless of the type of system or technique that the animator or graphic artist has at his disposal, he must be made to feel comfortable behind the computer's programming console. The planning of action sequences should be a relatively simple procedure. Staring at a console while dozens of buttons, dials, switches, and indicator lights stare back can be a distracting as well as a frustrating experience. Similarly, an elementary course in basic electronics should not be a prerequisite for the animator, whose major concern must be creativity. Unfortunately, a number of knowledgeable individuals with highly regarded positions in the industry refuse to accept computer-generated animation as anything more than a gimmick. At best, these diehards concede, "It's okay if you like electronic special effects." Producers and animators who have already gotten their feet wet merely smile on the way to the bank and reply, "Don't knock it 'til you've tried it." The pressured animator faced with a difficult deadline can no longer look up from behind a mountainous pile of drawings and offer the rather feeble excuse, "I'm doing my best—I'm not a machine, you know." Today, the "animator" could very well be a machine—analog or digital, on-line or off-line, static display or continuously variable—these animators need only be connected to the nearest electric outlet. The pencil in the guiding hand of the animator, however, continues to be the key that opens and activates the magic box known as the computer.

ANIMATION AND
ELECTRONIC IMAGING

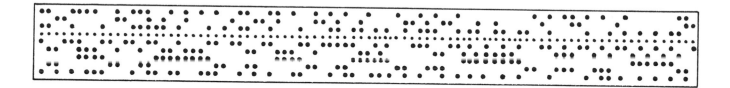

ANIMAC

Animac is an analog-computer system developed by Lee Harrison III, founder of the Computer Image Corporation. Referred to as the "magic box," this pioneer system for generating movement with static graphics not only helped set the pattern for more sophisticated computer systems but also served as a stepping-stone for the development of new techniques.

A surface-characteristics camera capable of distorting a planar surface to create dimensional effects was the next computerized-movement technique produced by Mr. Harrison. Operational in 1969, it led to the development of present-day analog-scan-conversion systems. See also **Caesar** and **Scanimate**.

BEFLIX

Developed by Dr. Kenneth Knowlton in 1964, **Beflix** is a computer-programming language used to perform drafting-type operations. The term is a corruption of Bell flicks for Bell Telephone Laboratories and flickers, a nickname for early motion pictures.

The output display is presented as 252 × 184 arrays of spots in different shades of gray, represented by the numbers 0 through 7. A complete map of the picture at every stage of programming is packed

in the computer's high-speed memory, three bits per spot. Coordinate numbers are assigned to each of the squares in the two-dimensional grid.

The Beflix programmer uses two levels of language. At the lowest and most detailed level of control the programmer directs scanners (**bugs**) to crawl about the two-dimensional array, reading the numbers that they pass over and optionally writing new numbers into these positions. On the higher level the programmer instructs the system in the preparation of graphics—basically geometrical shapes. The programmer can also direct operations on entire areas, giving instructions for copying, shifting, transliterating, zooming, dissolving, and filling areas bounded by previously drawn lines.

The Beflix system contains little mathematics; the operations it performs could also be done by hand. Sequences that are logically simple to describe but graphically or temporally intricate to perform are produced with considerably less effort and in a fraction of the time. Compare **Fortran IV Beflix**, **Explor**, **Fortran-coded Explor**, and **Tarps**.

The animation is achieved by applying the proper electronic signals to reshape the television raster. An image displayed on the raster will bend, squash, stretch, grow smaller or larger, and generally conform to the shape of the raster. A delta wave, for example, causes the raster to bend, along with any image displayed on it. Other wave forms, such as the sine, produce different types of distortion in the figure displayed on the raster. Squash and stretch effects are similarly controlled by manipulating the horizontal- and vertical-size controls. Acceleration and deceleration can be programmed to slow down or speed up an entire action or any part of it.

In place of the mountainous pile of drawings needed to produce a sequence of movement with conventional animation processes, the animator working with the Caesar system merely prepares a minimum number of sketches showing the cartoon character in full-face, profile, and three-quarter views, along with a small number of representative action poses. The component parts of the figure are rendered as high-contrast negatives, which are keyed for color by transparent gray overlays and fed into the computer's memory through a TV system. Any type of background may be used for the animation sequence—from paintings and photographs to specially prepared dimensional sets.

Since the animated figures perform in real time on the CRT display, their movement can be monitored, and changes in the action can be reprogrammed if necessary. As with conventional production techniques the animated figure's lip movements (lip sync) can be synchronized with a previously recorded sound track to add credibility to his antics. While it is possible to record the CRT display directly on film, the recommended procedure is to videotape it and subsequently transfer it to motion-picture film. See also **Animac** and **Scanimate**.

CAESAR

Caesar is an acronym for computer-animated episodes using single-axis rotation, a computerized-animation system. A descendent of Animac, the prototype animating computer designed by Lee Harrison III and his associates at Computer Image Corporation, Caesar is an analog device capable of functioning at high speeds. Programming and plotting key-frame sequences of animation with a digital-type computer is much too time-consuming compared with analog-type computer systems.

An animator familiar with the Caesar system could probably animate a scene in less time than it would take just to plan the action with conventional procedures.

With much of the tedium associated with the standard processes eliminated, the animator can be even more creative. The images, displayed on a high-resolution 945-line color-television screen, are quite good. While the system is not yet able to produce Disney-type cartoon animation, the results are far superior to many of the presentations shown on local television stations any Saturday or Sunday morning.

The system operates by dividing the component parts of a cartoon figure into a number of sections. Each of these components is displayed on an assigned part of the television raster: the head, for example, is displayed in one section, the body in another, the legs in a third segment, and the cartoon character's features in yet another part. Each of the parts of the animated figure can be controlled independently. Seated at the control console, the animator sees the various parts of the figure displayed on the monitor, manipulates the dials, and assembles a composite figure—a key frame. This frame is stored in the Caesar memory bank, and the animator prepares a second key frame for the continuing sequence of movement.

In conventional animation processes the key frames, or extreme drawings, are turned over to an inbetweener, who adds the number of drawings (specified by the animator) needed to complete that portion of the action. Caesar eliminates this time-consuming process. The animator specifies the number of frames that are to be inserted between the key frames; with his drawings logged into the computer's memory, the x- and y-coordinates representing the extreme positions of the action are the basis for calculations that produce the intermediate positions. For the computer this otherwise tedious task is performed in a matter of seconds. In many instances knowledgeable professionals have been unable to distinguish between interpolated computer-generated positions and inbetween drawings prepared by the animator's assistant.

CINETRON

Cinetron is a computerized recording system that creates a direct link between the animator or graphic artist and the animation stand or between the film editor and the optical printer. The computer is a self-contained unit that is capable of logging a programmer's instructions, making the necessary calculations, and filming the animation or opticals with great speed and mathematical precision.

The operation of the system is basically the same for both the animation stand and the optical printer. The main difference between the two systems lies in the command language, which contains routines that are basic functions of the optical printer—i.e., skip frame, multiple print, or color-head control. All Cinetron systems are controlled by on-line digital computers, with all functions monitored continuously by the central processing unit. The programmer or operator has the option of allowing the computer to control the entire operation or any portion of it. This type of control enables the cameraman to override previously programmed instructions and ad-lib at any point during the recording process.

In recording animation a series of photographs and/or static artwork is assembled as a unit and positioned on the compound table of the animation stand. The Cinetron system offers a welcome alternative to the tedious, time-consuming, and fatiguing frame-by-frame, stop-motion processes used for filming long sequences of limited animation. As in most of the computerized off-line or on-line animation recording systems, the animator or animation cameraman locates the start and stop positions for each piece of artwork in the sequence and programs instructions via a teletype, using industry terminology. The information fed into the computer's memory bank is translated into incremental steps that guide the camera through zooms and pans as well as cross-dissolves and fades. Accomplished at unbelievable speeds and with amazing accuracy, the moves are precisely repeatable.

If it is considered necessary, the action can be previewed by racking the camera over to its non-filming position and tracking the movement through the viewfinder. If the programmed action is fluid and meets script and storyboard requirements, the camera is racked back to its filming position, the computer takes over, and the animator and animation cameraman are free to take a coffee break. The odds are better than good that the filming process will be completed before the contents of the coffee cups are consumed.

The Cinetron-computer-software system requires no previous experience to operate. Instructions are given to the machine via a teletype or other input device such as a cathode-ray display with its own keyboard. The basic operation of the system is the same for any input method. The general form of the input data is: **literal identifier**, **numeric term or number**, **carriage return**, and **line feed**. The same general form for entering information is followed for each parameter given to the computer. The operator types a word that identifies to the computer the operation desired. The letters up, for example, indicate that the information that follows is part of the programming instruction used to describe an upward movement. Following this literal identifier is a comma. Each separate item on a line *must* be followed by a comma. It is used to separate different words or numbers given to the computer so that it can sort and store data correctly. A no-comma condition is diagnosed by the computer whenever the programmer types more than five consecutive letters, numbers, or spaces or if a number is entered before a comma in the first entry on the line. A comma is not necessary at the end of a line. Each line of data is terminated by depressing the carriage-return and the line-feed keys in turn, and the line is read into the computer's memory bank. The input data may be negated or erased by depressing the rub-out, carriage-return, and line-feed keys. By using the rub-out key comments may be typed at any time during data input to the computer, provided that a line of comments is terminated with a rub-out, carriage-return and line-feed sequence. Comments are particularly useful if the teletype-printed copy is to be stored with artwork for possible reshooting at a later date. Comments might include, for example, starting positions of a movement, frame numbers that differ from those given to the computer, identifying messages about the artwork, job numbers, or the cameraman's name.

Each entry in the allowable data line determines the computer's operations, based on the programmer's instructions. The first word, or the literal identifier, must be included within a list of words in the computer's memory bank. For the sake of simplicity and ease of operation only the first two letters of a word are validated by the computer. Since east, for example, is allowed by the computer, it will recognize any word beginning with the letters ea and will diagnose it as an instruction or series of instructions referring to an eastward movement of the animation stand's compound table. If the first two letters of a literal identifier contain a typographical error, the word will either be rejected or, if the error produces a different allowed word, be stored in the wrong area, with unpredictable results. If the first two letters of a line are invalid, the computer immediately types rejected after the carriage-return and line-feed steps and disregards all information appearing on that line. This procedure is a partial safeguard against accidentally reading comments as data if the programmer forgets the rub-out keys and against interpreting misspelled words as partial literal identifiers. The computer remains inoperative after the rejected message until the continue button is pressed.

If the programmer writes zero and depresses carriage return and line feed, the computer wipes out all previous information placed in its memory and in effect cleans the slate. Zero may be written at any time, but it must appear either before the end statement or as the first statement on a line. For example, the sequence zero, carriage return, line feed will cause all memory areas to be cleared, while the sequence east, zero, carriage return, line feed will *not* clear memory areas—instead, the letters "z-e-r-o" would be interpreted as part of an east/west movement. The result of that interpretation would be an "I'm confused" message. I'm confused or rewrite that, please, I'm confused indicate that a letter was written where a number was expected or vice versa. The computer will then remain inactive until the continue key is depressed, and the programmer is expected to retype the line correctly.

Shoot is the identifier for the target frame. The instructions curr, 10 (curr is the identifier for cur-

rent); shoot, 20; end cause the computer to expose 10 frames of film, with all programmed movements executed forward mathematically. Conversely, the instructions curr, 20; shoot, 10; end cause 10 frames to be exposed backwards, along with all previously programmed movement. Although the camera motor is in its reverse mode, the computer will not reverse the programmed direction of movement. Frame, typed as the first entry on a line, and followed by a carriage return and line feed, causes the current frame being used by the computer to be typed into the operation.

Curr(ent) is the identifier that tells the computer which frame it is to consider as the current frame with respect to the other instructions being typed. If the last current frame printed by the computer is the frame considered as current, no other programming instruction is needed. Under no circumstances may the current frame be set to ø unless the stop frame is set at a higher number.

Stop is the literal identifier for a preselected frame at which some manual operation is to be performed. In photographing animation, for example, the programmer might wish to add a superimposed title at some predetermined point. The stop-frame number indicates to the computer that every function required up to that frame must be performed before advancing to the next frame. The instruction stop, 1250 causes the computer to advance a frame count to 1250, make all programmed movements up to that frame, and stop before advancing to frame 1251. At that point the computer types the message manual action required current frame 1250 ready. When the continue button is depressed, the computer will again begin to run and will continue until all other operations are completed. New data may be entered by pressing revise and typing in the new data along with an end statement. End is typed by itself on a line and followed by carriage return and line feed to signal the end of the programming instructions. When end is typed, the computer immediately responds with ready and makes all needed preparations for carrying out the programmed instructions.

The following identifiers, accepted by the software program to designate motors and motor direction, are used to describe movement to the com-

88

puter: up, down, north, south, east, and west designate pan and tilt camera movements; clock designates movement of the rotation assembly in a clockwise direction; ccl describes a similar movement in a counterclockwise direction; 1w, 1e, 2w, 2e, 3w, and 3e identify each of the peg tracks used to register artwork on the animation stand's compound table, along with the specified direction of movement. North and south designations indicate the apparent movement of the centerline of the camera in the programmed direction: for example, north, 100 would cause the centerline of the camera to appear to have moved northward 1″. This apparent movement northward would require a southward movement of the compound table, since the camera remains in a stationary position throughout the movement.

The **parameter** format for describing a **movement routine** is as follows: **literal identifier**, **distance to be moved** in hundredths of an inch, **beginning frame number** indicated on the counter for the beginning of the movement, **end frame number** showing on the counter for the end of the movement, **frame count** to be used to taper the movement, and the **literal indication** in, out, or both. If in is typed in this position, the movement will be tapered in the number of frames indicated at a constant speed and will be continued at that speed until the last frame of the movement. If out is typed, the movement will begin at speed and taper out. If both is typed, the movement will taper in and out in the number of frames indicated in the taper section. The distance of the movement is stated in hundredths of an inch and is given to the computer as a whole number. Fractions or decimals are not allowable inputs at any time except as comments.

Tapers are plotted by the computer on an inverse-sine relationship. The shape of the sine, which controls the shape of the taper, can easily be modified by the programmer to meet specific requirements. The greater the number of frames in the taper, the greater the distance covered per frame after the acceleration is over. For example, a 35-frame movement with a 5-frame taper at both ends will move more gradually through the 15 center frames than the same movement with 10-frame tapers at both ends.

An exponential zoom appears to travel at a constant speed with respect to the field being photographed. It is indicated as follows: up/down top position, start frame, stop frame, bottom position, exp. The top and bottom positions indicated are the veeder counter readings for the top and bottom positions of the movement. Exponential zooms are useful for matching screen speeds of different sized artwork and for matching zooms to perspective drawn in artwork. Exponential off-center movements will match speeds with respect to field size and number of frames, as do exponential center-zoom movements.

A typical off-center zoom may be accomplished in the following manner. In this example all three axes are used. Assuming that the move is to begin after frame 200, a move length of 275 frames and a taper of 14 frames at both ends is specified. In the first step the film in the camera is exposed through frame 200 (i.e., 200 shows on the camera counter). In the next step the cameraman toggles in 275 on the total-frame thumb-wheel switches, 14 on the taper-length thumb-wheel switches, and selects both on the taper selector. The compound table is moved to the desired starting position, and the enter-start switch is set. The compound table is then moved to the end position of the zoom effect and the enter-stop switch is set. In order to view the move without exposing the film in the camera, the cameraman throws the system into the disable mode and pushes the run button. Each axis will automatically return to the start position. If the simulate switch is in the continuous mode, the system will step through the move. If the cameraman wishes to single-cycle at any point, simulate is switched out of continuous. The single-cycle mode is initiated by pushing the same switch. Returning to continuous-simulate causes the system to complete the move automatically. If the previewed move is satisfactory, the simulate mode is negated, the enable switch is thrown, and the run button is depressed. The system will return the compound table to the starting position and carry out the move, with the camera exposing automatically through the 275 frames of the move, including the 14-frame taper at both ends of the effect. When the move is completed, the camera counter will read 475, and each axis will be in its stop position.

If the cameraman wants to stop the system at any point during the run to change artwork, for example, he simply throws the system into disable. The cameraman may then single-cycle to the desired frame, using the camera-shoot button. If the program must be stopped in order to erase the move, the emergency-stop switch is activated. The system's design allows a shoot-to-frame (up to 999 frames) to be programmed simply by selecting the length of the move, pushing the enter-start and enter-stop switches, and initiating the exposure process with the run command. The procedure is applicable for both the animation stand and the optical printer.

The **write** routine is used if the cameraman wishes to see the effect of a movement without actually moving the compound table or any other part of the animation camera. The routine is used to write the cumulative current position of a programmed movement. The numbers displayed (repeated) by the computer indicate the current position relative to the zero starting point in thousandths of an inch. All data parameters normally used in operating the camera must be present for a write routine: for example, there must be a frame count, the frame count must be valid with respect to starting and stopping frames, and all programmed movement must take place between those specific points. The computer will not move the compound table or activate motors, either positional or camera, while the write routine is in affect. To negate the write routine, the cameraman must write zero or type the instruction expose, which resets all data areas and prepares the computer for all subsequent operations connected with the shooting of the written plot.

The **sequence** routine allows the cameraman to shoot a large number of sequential pieces of artwork in a very efficient manner. The input of information for the routine is as follows: sequ (number of cels or pieces of art plus 1), number of exposures per piece of art, number of repeats of cels and exposures. The input for the exposure of five cels, with three exposures per cel and thirty repeats, would be written as sequ, 6, 3, 30, followed by depressing the carriage-return and line-feed keys. Frame data must also be included. With the typing of the end statement the computer will ready, and the camera-

man positions the first piece of artwork on the registration pegs of the animation stand's compound table. Pressing continue, the computer will shoot the first piece of artwork along with any programmed movements in each frame position indicated by the call. It would in this case shoot the first piece of art at frames 1, 6, 11, and so on, with the artwork in its proper position at each frame in the sequence. The shutter would be capped for the frames of film that are not exposed during that particular operation. After the cameraman receives the message next cel please manual action required current frame xxx (frame number shows on camera-frame counter) ready from the computer, he removes the first piece of art, replaces it with the second, press continue, and repeats the procedure until the message received from the computer is current frame xxx (number shows on frame counter) data? In order to conclude a sequencing operation, the cameraman or programmer must type zero before entering new data.

The **hookup** routine controls the movement of the peg tracks, used for registering artwork, on the animation stand's compound table. The routine enables the cameraman to hookup, or repeat, the picture information in a piece of artwork, usually a background, at regular intervals. If a pan background has two precisely repeatable positions 27″ apart and if it must be hooked up five times during the filming of the sequence, it would have to be programmed to move 13,500/100″, or 135″. The cameraman would type 13,500 as the travel distance for the peg track on which the background is registered, in this case peg track 3, and the instruction H3, 2700, followed by carriage return and line feed. This would indicate to the computer that every time a position equal to 2,700 (100″ from its beginning point) is reached, the artwork must be returned to the hookup position before the panning action can continue.

The largest number that the computer program can handle, 32767, has only five digits. When it encounters five consecutive letters or spaces not separated by commas, the computer types the message no comma, and the programmer must consider all data entered on that line as invalid. He must reset the line to indicate the movement desired in a way that the computer will accept. Failure to retype the line with valid instructions produces unpredictable and usually undesirable results. As in the case of the rejected message, the continue button must be pressed before the line can be retyped.

If accurate is followed by carriage return and line feed and is the only data entered on a given line, the computer sets up the appropriate action so that each movement motor is ready to operate in a counterclockwise direction. An eastward movement, for example, requires a clockwise rotation of the east/west motor. The **accurate** routine causes the east/west motor to run past the designated starting point and return in a counterclockwise direction to the same point. This routine enables all chains, drives, and lead screws to be approached from the same side regardless of the direction of movement involved in the operation. The accurate routine is required for most operations in which multiple runs or matte shots make critical registration a necessity.

If again is typed into the computer, followed by carriage return and line feed, as the first and only entry on a line, the computer repeats the movements stated in the previous call for data. After again is typed, the current frame is subtracted from all counts relative to the starting frame. If a new movement is to be added, that information can be included before again is retyped. Followed by carriage return and line feed, the **again** routine causes the computer to ready immediately, so it can be added only after all other instructions pertaining to the movement have been typed. If reverse is typed as the only entry on a line and is followed by carriage return and line feed, an operation similar to the again routine is performed. The **reverse** routine repeats the previously programmed routine but with the motors reversed: a fade-in, for example, becomes a fade-out.

The **rotate** routine causes the last movement or set of instructions to be repeated backwards—both physically and mathematically: an east taper or ease-out becomes a west taper or ease-in. The count at the end of the movement is the same as the starting count in the previous movement. The instruction curr, 1; shoot, 10; end causes the camera to expose 10 frames; rotate, combined with the manual reversal of the camera, causes the camera

to expose backwards from frames 10 to 1. At that point the current frame held by the computer would be 1.

Black, followed by carriage return and line feed, causes the camera shutter to close. If the shutter is already in the closed position, it will remain closed. The typed command open causes the camera shutter to open. Index, typed on a single line by itself and followed by carriage return and line feed, causes all motors (except the camera motor) to return to the last programmed position. Similarly, lock instructs the computer to hold the current compound-table positions for the index command that follows. Audition instructs the computer to carry out the programmed action without shooting. It is used to preview a movement and is removed by typing zero or expose. The instruction to expose, typed as a single entry followed by carriage return and line feed, negates the audition or write mode. Typed as a single entry and followed by carriage return and line feed, wait suspends the read operation and allows the cameraman to check positions and modes after entering data but before typing end.

A programmed movement that exceeds the equipment's capability or preset parameters is sensed by the computer on an input line to memory. If a **limit** is proposed or reached in any way other than by the action of the computer, an indicator light will remain on as a signal to the cameraman. It is then necessary for the cameraman to move the limit manually before any operation (including write) may be performed by the computer.

The computer considers a command to fade or dissolve in or out as a priority command. If the instruction to fade in is encountered while the shutter is already open, the computer will close the shutter and fade in. The reverse action would be taken if the fade-out command occurred while the shutter was already in its closed position. The instructions cap and uncap cause the shutter to be opened or closed even if they duplicate the current mode. The opening or closing corrective action is taken by the computer on the first frame in the fade or dissolve description: for example, shoot one frame, close shutter, fade in, finish. A fade in, starting at frame 10 and lasting 16 frames, would be entered as follows: fade, 10, 16 in. Since the computer counts each frame as it is advanced, a 10-frame dissolve from frames 10 to 20 would be typed as follows: diss, 10, 10, out; diss, 20, 10, in; stop, 21; curr, 1; shoot, 30; end. The film-stock-compensated logarithmic fade makes it possible to shoot a 2,000-frame fade with only 1 frame of black. Each of the other frames will change in brightness to match the number of frames in the fade. Similarly, the dissolves are sine-shaped and do not display the "pop" normally expected from mechanical units. In order to complete a cross-dissolve, the film in the camera must be returned to the frame in which the effect began with the shutter in the closed position. Typing motor causes the camera-drive motor to reverse direction. If it is in a forward mode, it will reverse itself; if the camera is already running in reverse, it will change direction and proceed in a forward mode.

The **supervisor** mode allows the computer to advance the program one frame at a time while examining a tape or keyboard command that may either change the operation of a program or the access random operations on the equipment being controlled (cap, uncap, forward, reverse, etc.). In order to initiate the mode, supervise must be typed in the data section of the program on its own line and terminated with the standard carriage return and line feed. The mode may be negated by typing zero or abort. The supervisor mode normally operates from paper-tape instructions that are prepared in advance. It allows complete preplanning of a sequence—plotting movement, setting the equipment controls, and subsequent photography, including the preparation of an "exposure sheet," the control tape. After the photographic process begins, the cameraman must change to artwork—the only manual contact with the computer-controlled equipment.

Since the first Cinetron system went into operation in 1967, new systems have been developed from the animator's and the cameraman's points of view. The flexibility of design allows a changeover to computerization at studios using conventional equipment with a minimum of downtime. The software system is designed to be operated by people who have no previous experience with computers.

CONTINUOUSLY VARIABLE DISPLAY IMAGES

This method for generating and subsequently recording images—known as **continuously variable display images**, dynamic computer-generated animation imaging, or dynamic computer imaging—is made possible by the unique capabilities of the analog computer. Analog systems, in addition to their mathematical proficiency, are able to differentiate and integrate the continuous input of image-forming bits on the basis of changes in the magnitude of the signals. The input signals used for generating an image can be altered by the analog's dissimilar circuit elements to produce changes in both the shape and position of the displayed images.

A simple diagonal line formed by the CRT's electron-beam trace can be altered in any number of ways by an analog circuit that is used to modify the magnitude of the input signal. Shifting the phase of the CRT's horizontal- and/or vertical-deflection elements can cause the diagonal line to reform in the shape of a circle or continue through a series of elliptical shapes before returning to its original position on the display—at the touch of a button or turn of a dial. The increased capacity of complex analog computers can be used to manipulate the simplest of input signals to produce the most complicated graphic displays.

In a variation of the technique a scanning device is used to convert previously prepared graphics into appropriate analog signals. Functioning as a camera, the scanner "exposes" the artwork in a manner similar to the scanning of a scene by a conventional TV camera. Displayed on a CRT, the artwork can be manipulated through circuitry that alters frequency, amplitude, wave form, and other image-controlling variables. In either of the two systems—generating image-forming signals with analog circuits or analog signals with image-scanning devices—the images are photographed in real time with conventional motion-picture cameras operating at the standard 24-frames-per-second rate.

The technique enjoys a number of distinct advantages over other image-generating and recording techniques. Artwork can be prepared and programmed more rapidly and with considerably less effort than is required for the static-display-imaging system. Motion sequences generated with these imaging techniques can be previewed on monitors as they are being created; if the dry run indicates a need for revisions, changes can be made electronically by merely turning a dial or flipping a switch, and the output recorded on film in real time.

The best features of the digital and analog computers have already been combined into a hybrid. The static and the variable imaging techniques complement each other, and, while each has been used successfully in commercial production, the best features of both will be used as a stepping-stone towards the development of new techniques. Compare static computer-generated animation imaging.

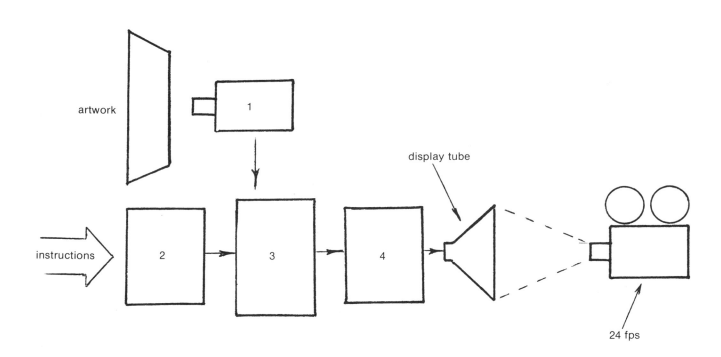

7-1. Continuously variable display imaging. In this system the image is scanned by a television camera, and all the component parts are stored in the computer's memory bank. The programmer at the control console can manipulate the image with pushbuttons and dials that control the analog circuit's image-altering elements. The units of the system are: (1) television camera, (2) control panel, (3) analog computer, (4) display hardware.

DATAVISION

The 3M Company's **Datavision** Model D-2000 video character generator emphasizes high-resolution characters and versatility. Introduced by the Mincom Division, the video character generator features internal video mixing and a built-in audio-recording storage interface. It is capable of displaying 62 alphanumeric characters and symbols in a ten-row 24-character-per-row format, and resolution for each of the generated characters is equal to a maximum of 1,120 video elements. The unit's random-access memory can store as many as 40 rows of data. The character generator's built-in audio interface provides for storing data on most voice-graded audio recorders. The unit also accepts video input, so the video portion can be mixed with the generated characters, all of which can be positioned in one or two rows at the top, center, or bottom of the image area.

The Model 5120 Video Outliner and the Keyer-Colorizer are two recent additions to the company's electronic-video-effects-equipment catalog. Also introduced by the Mincom Division, the outliner enables shadow edging or borders to be included around generated characters or design elements. The unit can provide either black or white edging or outline only the transparent characters. The Keyer-Colorizer Model 5110 allows color to be added to camera-produced graphic material. Three separate video inputs provide pushbutton selection of any combination of the three and full nonadditive mix of any two or all three.

EDUCATION AND RESEARCH, COMPUTERIZED

The availability of computer-controlled systems for generating graphics and producing movement has enabled educators and researchers to process and present great quantities of visual information, both real and simulated, at incredible speeds. The mathematical capabilities of these computer systems provide the means for illustrating solutions to complex problems associated with space technology with the same amount of effort that is required to animate corporation logos and abstract patterns for television commercials. Computer-produced movies are playing an increasing role in education and research. A number of computer films produced at Bell Telephone Laboratories since 1963 demonstrate the dynamic, graphical power of computers and automatic film-recording equipment. Movies made by computer are a particularly significant adjunct to education and scientific investigation in fields amenable to mathematical and logical treatment and in which results can or should be visualized.

An electronic microfilm recorder can plot points and draw lines a million times faster than a human draftsman. This machine and the electronic computer that controls it make feasible movies that up to now have been prohibitively intricate, time-consuming, and expensive to draw and film. The **microfilm recorder** consists of a display tube and a camera and understands only simple instructions such as to advance the film, to display a spot or alphabetic character at specified coordinates, or to draw a straight line from one point to another. Though this repertoire is simple, the machine can compose complicated pictures or series of pictures from a large number of basic elements. It can draw as many as 100,000 points, lines, or characters per second. This film-exposing device is therefore fast enough to turn out in a matter of seconds a television-quality image consisting of a fine mosaic of closely spaced spots or to produce simple line drawings at the rate of several frames per second. The important question is not whether sufficiently detailed pictures can be produced quickly enough but whether sufficiently powerful and useful programming languages and techniques can be devised and whether the resulting computation times would be reasonable. The answer to this question is definitely yes. Useful techniques can be and have been developed, and computation times for most purposes are quite reasonable. This is demonstrated by the Bell Telephone computer-produced movies mentioned above. Grouped under the two headings of education and research, these films were produced by programming the IBM 7094 computer to run the Stromberg-Carlson 4020 microfilm recorder. The programming languages used were Fortran, with microfilm commands added, and Beflix, the mosaic-picture system developed by Dr. Kenneth Knowlton.

In these filming projects the computer played two distinct roles: that of a high-powered drafting machine and, particularly in scientific and mathematical areas, that of a calculating machine that determines the consequences of mathematical and logical statements. In the latter role the computer typically accepts a description of a hypothetical system, determines its successive states by following differential equations or other supplied laws, and uses its drawing capabilities to render a series of views of the resulting events. One of the films is a straightforward exposition of the Beflix movie system, produced by the system itself. It demonstrates the output of the system as well as presenting it from the programmer's point of view. In Beflix pictures are thought of as 252-×-184 arrays of spots in different shades of gray or as numbers from 0 to 7. The computer keeps a complete current map of the picture packed in its high-speed memory. Like other Beflix-made films, this film was produced at a cost of about $500 per minute, which is approximately the same rate as that of the most economical animated film produced with conventional processes. About one-half of this cost can be attributed to the programming, about one-third to computer time, and the balance to optical printing and recording the accompanying sound track.

The Beflix programmer uses two levels of language. At the lowest and most detailed level of control he directs scanners to crawl about the two-dimensional array, reading the numbers on which they are located and optionally writing new numbers into these positions. On the higher level the programmer instructs the system (with its own scanners that do most of the work) to draw lines, arcs, letters, and other curves. He can also invoke operations on entire areas and give instructions for copying, shifting, transliterating, zooming, dissolving, and filling areas bounded by previously drawn lines. The Beflix system contains little mathematics, and operations that could be done by hand are best left to the computer, which can perform them more easily in a fraction of the time. This is especially true of sequences that are logically simple to describe but graphically or temporally intricate.

A basic lesson in electrical engineering is offered

in the film *Harmonic Phasors*, produced at Bell Labs by Prof. William H. Huggins of John Hopkins University and Prof. Donald D. Weiner of Syracuse University. The computer-generated film illustrates the composition of complicated periodic wave forms by adding projections onto an axis of rotating vectors, or phasors. The subject matter is presented

7-2. Scene from a film about the Beflix movie system produced by Beflix. Depicted are a card deck of instructions, a computer, and a tape, on which complete spot-by-spot picture descriptions are written and later read by a microfilm printer. Each picture is actually a 252-×-184 mosaic of typed characters—degree signs and apostrophes in this case.

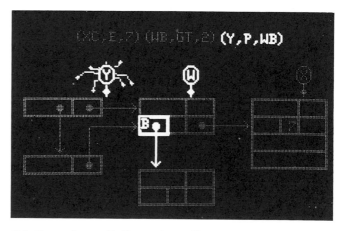

7-3. Scene from a Beflix-produced film showing the function of the bug as it moves from one part of the grid to another in accordance with the programmed instructions.

in dynamic and graphical symbolism—it is much more vivid than blackboard illustration. The film is a good example of an essentially mathematical situation that the computer can handle easily, providing smooth and accurate animation that would normally be quite difficult to achieve. The film was programmed in Huggins' own macrolanguage for dealing with rotating vectors and related material, a language that has since been used for other films produced by the National Committee for Electrical Engineering Films, of which Huggins is a member.

The mathematical role of the computer is even more obvious in F. W. Sinden's educational film on Newton's law (f = ma) as it governs the motion of heavenly bodies under the law of gravity (the inverse-square law) and as it would govern these motions under other hypothetical laws. One scene of the sequence shows the motions of two bodies that would result from a mutual attraction varying directly as the cube of the separating distance. From the filmmaker's point of view Sinden's film demonstrates how a pictorially wide variety of scenes can be produced by feeding a few subroutines in turn with the various parameters corresponding to the

7-4. Scene from F. W. Sinden's *Force, Mass, and Motion*, a film that demonstrates Newton's law of motion (f = ma) for different central forces. Shown here are paths of bodies mutually attracted by a force that varies directly as the cube of the separating distance. The programming language was Fortran.

different situations to be portrayed. From the pedagogical point of view its smooth animation and mathematical precision make it an elegant and aesthetically beautiful lesson in physics. Its total effect cannot be matched by any amount of writing at the blackboard or by any physically realizable demonstration equipment. This suggests that the computer is the perfect physics laboratory, for it can abstract, idealize, or revise the laws of nature to a far greater extent than any other teaching aids or experimental equipment.

In scientific research the most obvious moviemaking function of the computer is simulation, with results presented graphically and dynamically. A simpler job is the straightforward presentation of raw or only slightly cooked data in a form more immediately comprehensible than graphs or lists of numbers. Other computer products are dynamic displays, created according to mathematical specifications, for a variety of purposes: as stimuli for experiments in perception, as a study of the mathematics involved, or as an exploration into artistic and aesthetic possibilities.

The first computer film produced at Bell Labs, made in 1963 by E. E. Zajac, simulated the motion of a communications satellite. The problem under study was the orientation and stabilization of the satellite so that one end and the antenna constantly pointed toward the earth. The orienting force was the gravity-gradient torque, which results if one end of the elongated satellite is slightly closer to the earth, and stability was achieved by viscous-coupled gyros that damped out oscillatory motion. The simulation required numerical integration of complicated differential equations. Results were presented as perspective drawings showing the position and orientation of the satellite and its gyros as a function of time. Several frames of this film show the earth, the box-shaped satellite in several positions, and a multiply exposed clock used to count orbits. This film was a triumph, not because it helped Zajac in his study but because it enabled him to explain his results to someone not thoroughly familiar with the problem.

An example of a computer-generated production that did significantly help the investigator is a film produced by D. E. McCumber that shows the instantaneous-electric-field strength as a function of distance through a semiconductor. Under investigation in this film was the Gunn-effect instability, in which electric-field spikes repeatedly arise

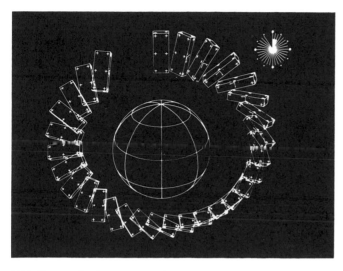

7-5. Composite from several frames of E. E. Zajac's film on satellite orientation by means of gravity-gradient torque. The lines on the side of the box indicate instantaneous orientations of internal gyros; the multiply exposed clock at upper right counts the orbits.

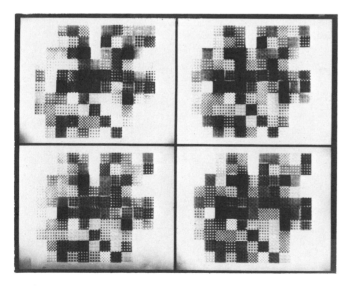

7-6. Sequence taken at 10-second intervals from R. A. Semplak's film depicting the instantaneous rainfall rate in a 60-square-mile area. Light areas signify no rain; darker patches, heavy rainfall.

and propagate under certain configurations and applied voltages. This particular film and others like it revealed to McCumber and his associates the precise nature of the spike initiation implicit in but not immediately apparent from the mathematical description of the phenomenon.

An unusual presentation of raw data is shown in a film by R. D. Semplak that depicts a rainstorm in a 7-mile-square section of New Jersey. Semplak's study attempted to associate some microwave-communications problems with certain shower forms and to judge the feasibility of selecting alternate microwave-transmission routes when rain is interfering. The computer made rainfall over a wide area visible without sophisticated mathematical treatment by taking the data from a large number of rain gauges and displaying the measurements instantaneously.

The realm of mathematically specified pictures is illustrated in a film produced by B. Julesz and C. Bosche. The production contains a large number of sequences based upon noise from a pseudo-random number generator with certain constraints of symmetry, periodicity, and other spatial and temporal regularities. The area of study is human vision and perception, and the computer provides the means for producing hundreds of displays that are pure in the sense that they are almost completely devoid of familiar patterns. The investigator can use these patterns, for example, to determine conditions under which subjects see three-dimensional effects independently of any knowledge of the object seen. Without the computer and electronic-display equipment it would be impossible to prepare these experimental situations, because each frame of film contains tens of thousands of black and white spots.

Another mathematical construction is shown in four successive stereo pairs from a film by A. M. Noll. They are views of a four-dimensional hypercube projected mathematically into three dimensions and twice into two dimensions—onto slightly different picture planes for the left and the right eye. One form of this film has side-by-side images for cross-eyed viewing; another version is designed for a special projector and the familiar 3-D crossed-polaroid glasses. This demonstration is part of an interesting but so far unsuccessful attempt to give people an intuitive feeling for four-dimensional objects by exposing them to a variety of these objects manipulated in four-dimensional space by such operations as rotation about their four mutually perpendicular axes. Noll's stereo-display facilities have been used for a variety of other scientific and artistic purposes, including such diverse applications as experimental documentation for choreography and the depiction of part of the hearing process in the inner ear.

Two films, made mainly for the fun involved in producing them, demonstrate rather broad areas of humor and art in which the computer plays its own particular part. The first, by R. N. Shepard and E. E. Zajac, shows a computer-animated ball bouncing ever upward on the Penrose-and-Penrose staircase. The accompanying sound is a cycle of 12 computer-generated tones that seem like an endless sequence of ever-increasing pitch. The second is a short film made in connection with Expo 67 by Dr. Kenneth Knowlton and Stan Vanderbeek, a

producer of experimental and avant-garde films. It is a visual play upon the theme of the fair, *Man and his World*, in several languages. It was produced by programming a special set of macroinstructions that in turn were written in terms of Beflix operations.

Significant work in computer-animation-gener-

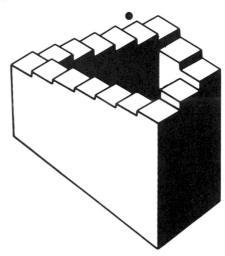

7-8. Frame from *A Pair of Paradoxes*, in which a ball programmed by E. E. Zajac bounces up an illusory staircase, accompanied by R. N. Shepard's cycling tone sequence of apparently ever-increasing pitch.

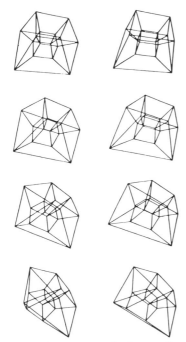

7-7. Four successive stereo pairs from a film by A. M. Noll designed for stereo viewing. Portrayed here is a rotating four-dimensional hypercube projected into three dimensions and twice onto two picture planes, producing two slightly different views for the left and right eyes.

7-9. Scene from a short film by S. Vanderbeek and K. C. Knowlton produced for Expo 67. This movie was programmed in a higher-level set of macroinstructions, defined in terms of Beflix instructions, and specifically designed to manipulate textual information in intricate ways.

ating and -recording techniques for educational purposes, scientific research, and commercial television is being done today in production studios, industrial plants, college classrooms, and experimental laboratories in just about every part of the country. At the Boeing Company, for example, computer projects are concerned mainly with aircraft design and cockpit visibility. The Las Alamos Scientific Laboratory project is devoted largely to simulation and visual presentation of fluid phenomena. The Lawrence Radiation Laboratory (LRL) has produced a variety of color films depicting simulated weather conditions, explosions, and other physical phenomena. Another group is actively supervising the production of a number of films at several universities. The National Committee for Electrical Engineering Films (NCEEF), chaired by Prof. John Brainerd at the University of Pennsylvania, has found computer animation very helpful for sections of their films that require visualization of time-varying vector fields or the presentation of complicated mathematical concepts. Impressed with the wider applicability of computer animation, the NCEEF has also put out a 20-minute sampler of films made at Bell Telephone Laboratories and elsewhere.

The computer and the automatic film recorder, because of their high calculation and display speeds, make it possible to produce films that would be far too expensive or difficult to produce with conventional animation processes. Costs for the films have now fallen to the range of $200 to $2,000 per minute; the cost for corresponding hand-animated films would be at least twice as much for less complicated productions and entirely unrealistic in other cases.

The computer offers the further advantage of few intermediaries and few delays between the producer and the filmmaking mechanism, speeding up the overall process and minimizing communication problems. In the films discussed in this section the producers were also the instructors or researchers rather than hired animators. The speed, ease, and economy of computer animation permit the moviemaker to film a scene several times—producing a whole family of clips—and to select the most graphically appealing and representative one, a luxury never before possible.

EXPLOR

Explor, a computer language developed by Dr. Kenneth Knowlton of Bell Telephone Laboratories, is an acronym for explicitly provided 2-D patterns, local-neighborhood operations, and randomness. It builds and stores pictures in **raster-scan** format, with an eight-bit number retaining information pertaining to each (x, y) position. It is a two-dimensional language only in that one of the inputs to the program may be cards on which successive rows of two-dimensional input patterns are punched. The language is useful not only in providing the computer novice with graphic output but also as a vehicle for introducing many other basic computational functions: algorithms, nested loops, sorting, heuristics and search, cellular automata, Monte Carlo calculations, and finite-state machines. It has proven to be effective in depicting results of simulations in natural (i.e., crystal growth) and hypothetical (e.g., cellular automata) situations and in producing a wide variety of designs.

Mini-Explor is a simplified version of the Explor system. It is coded in 430 lines of Fortran, which can run on most 16-bit-word (or larger) machines with at least 8K of core storage. By means of lineprinter or teletype output such an implementation can now produce almost as rich a variety of results as much larger versions.

Unless otherwise modified, output is by means of write statements that cause up to three-times-overprinted output on the machine's line printer or teletype, yielding four effective shades of the gray scale. The internal image is retained in raster-scan format and consists of 140 lines of 140 spots each, packed as seven two-bit picture cells per machine word. Most operations involve unpacking and repacking sections of the total image; an implementer may find it desirable to recode in machine language the low-level outlines that perform these tasks.

From the programmer's point of view the system consists of nine Fortran-callable **functions** and **subroutines**. In order to use the routines effectively, literacy in the following Fortran functions is necessary. (see Fortran IV Beflix): subroutine calls; gø tø's; integer variables; arithmetic-assignment statements; operators + – * / **; dø loops; logical IF statements; connectives .and. .ør. .nøt.; built-in functions mino, maxo, mod, iabs, isign; one-, two-, and three-dimensional arrays. Explor functions are: num (x,y) and ne (min,max). Subroutines are: call shøw (x, y, w, h), call put (x, y, n), call put4 (x, y, n), call put 16 (x, y, n1, n2, n3, n4), call chanj (x, y, w, h, % rule), call løcøp (x, y, w, h, %, many, nabors, these, rule), and call cømbn (x, y, w, h, %, xf, yf, orientation, r0, r1, r2, r3).

All parameters used in subroutine calls are integers (i.e., with values 0, 1, 2, 3, 4 . . .); their values may be negative. It is advisable to make the first line of every program, if it is accepted by the local Fortran compiler, the following: implicit integer (A-Z). The user imagines the internally stored picture as a 140-×-140 **array** of picture cells, each holding a digit 0, 1, 2, or 3 and addressed in terms of their x, y coordinates. At the beginning of a program all cells are filled with zeros. Subroutines dealing with rectangular areas include in their descriptions the **dummy parameters** (x, y, w, h, %, . . .), which have these meanings: x is the x-coordinate of the center of the rectangle (or 1/2 cell left of center if width is an even number); y is the y-coordinate of the center of the rectangle (or 1/2 cell below the center if height is even); w is its width; h is its height; and % is an integer 1 to 100 stating the approximate percentage of cells actually to be treated on a pseudorandom basis (100 indicates that all of the cells are to be processed). Compare Beflix, Fortran IV Beflix, Fortran-coded Explor, and Tarps.

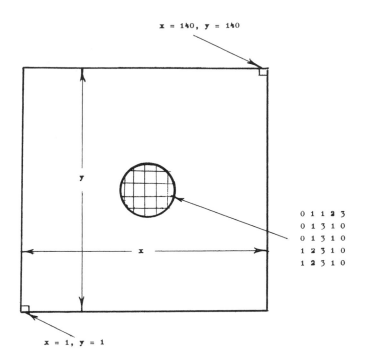

7-10. The 140-×-140 Explor grid.

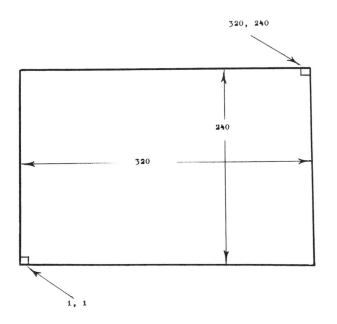

7-11. The Fortran-coded Explor grid.

100

FORTRAN-CODED EXPLOR

Fortran-coded Explor is a simplified version of the Explor language, also developed by Dr. Kenneth Knowlton of Bell Telephone Laboratories. It is used to teach computer graphics and art at the college level. The language is based on **alphamerics**. The "surface" inside the computer is a two-dimensional grid, each cell of which contains at any time a nonnegative number from 0 through 255. For card input and nontransliterated card or printer output, the first 63 of these numbers are expressed by single characters:

```
0123456789abcdefghijklmnopqrstuvwxyzØ.(+$*)-/,_="<|&;-|%>?:#@'¢!
|         |         |         |     | |         |         |
0         10        20        30    36 40        50        60
                                  (blank)
```

Numbers greater than 62 cannot be entered in a simple way via cards, though it is possible to punch all of the "characters" and thus generate any number from 0 through 255 by multipunching card columns. All variables unless otherwise stated are integers; in most cases in which it seems meaningful their values may be negative. The dimensions of the grid are 240-×-320 squares.

Several subroutines perform operations on rectangular subareas; the first five dummy parameters in descriptions of these subroutines are written with the symbols x, y, w, h, %, . . ., which have these meanings: x is the x-coordinate of the center of the rectangle (or 1/2 square left of the center if width is even); y is the y-coordinate of the center of the rectangle (or 1/2 square below the center if height is even); w is the width of the rectangle; h is the height of the rectangle; % is the square-by-square percentage probability of actually performing the operation, normally an integer from 1 to 100. Negative numbers and zero are taken to mean no operation; numbers greater than 100 are taken, as 100, to indicate that all squares are treated. Illegal or questionable values of parameters defining such rectangular areas elicit warning comments in the printout, generally at decreasing frequency. The following conditions result in no operation: defined rectangle entirely off the surface or zero or negative height or width. Compare Beflix, Fortran IV Beflix, Explor, and Tarps.

FORTRAN IV BEFLIX

The **Fortran IV Beflix** computer-programming language combines the area-filling gray-scale facilities of the Beflix language with the mathematical capabilities of the Fortran system. The Fortran mathematics used for generating graphics (the construction of geometric shapes, for example) are incorporated within the framework of Beflix. The combination of the two languages, revised and programmed in Fortran IV, makes the Beflix animated-movie language far more machine-independent.

The two-dimensional internal picture storage In Fortran IV Beflix is similar to the storage arrangement in the original Beflix system. The two-dimensional grid is used for creating and manipulating **packed arrays**—groupings of picture elements—in a raster-scan representation. Each square in the internal grid is identified by an x, y integer pair. Provision for packing, unpacking, and shifting machine words, as well as the subroutine for the output of the array, are the four major machine-language subroutines required in any installation using this programming language. The configuration of computer and output-display hardware is optional and is therefore expected to vary with the individual installation. Compare Beflix, Explor, Fortran-coded Explor, and Tarps.

The most frequently used programming entries are defined below. These calls, describing equipment functions, initiate specific routines and operational programs that are controlled by the computer.

Grid Operations

window A rectangular area represented by the internal two-dimensional grid used for programming. The window is depicted in the positive quadrant of the plane, with each square identified by an (x, y) integer pair. The area represented may be changed by the programming routine call window (x-min, y-min), in which x-min and y-min become the leftmost column and the bottommost row for the new window.

Bug Operations

place One of the grid operations in the Fortran IV Beflix programming language. Base pointers or scanners—**bugs**—symbolize the movement of picture elements from one coordinate position in the grid to another. Before a bug is used, however, it must be initialized or identified at coordinates by the routine call place (bug, x, y). It may then be moved to a new (x, y) coordinate by the routine call møve (bug, tø, bug 2). Bug positions cannot be changed by any subroutine other than møve.

Drafting Operations

rect One of the drawing operations in the Fortran language. The routine call rect draws a rectangle (with number n or with random numbers specified by R) of indicated width, with right and left the extreme side boundaries in which the vertical lines are drawn, and top and bottom the extreme boundaries for drawing the horizontal lines. The parameters for the routines are call rect (right, top, left, bottom, $\frac{R}{n}$, width).

line Call line $(x_1, y_1, x_2, y_2, n, \text{width}, \text{frames}, \text{speed})$ are the parameters for the routine. This drafting operation draws a straight line by a sequence of horizontal, vertical, and 45° steps from (x_1, y_1) to (x_2, y_2).

arc One of the drafting routines in Fortran programming. Call arc $(x_1, y_1, x_c, y_c, \frac{cw}{ccw}, n, \text{width}, \text{x-limit}, \text{y-limit}, \text{frames}, \text{speed})$ describes an arc beginning at (x_1, y_1) with its center at (x_c, y_c), proceeding either clockwise (cw) or counterclockwise (ccw). The arc stops when the line reaches a specified x- or y-limit or when a complete circle has been drawn.

trace This routine is used to trace an arbitrary curve in response to incremental directions and amounts. The programmed directions are expressed by the parameters call trace $(x_1, y_1, \text{string}, n, \text{width}, \text{frames}, \text{speed})$.

type This routine is used to type dynamic alphameric characters. In programming the lower-left corner of the first letter in the string of characters is assigned the location (x_1, y_1). In addition to the letters of the alphabet the text may include all of the numbers and symbols that the typewriter keyboard can produce. The parameters for the routine are call type $(x_1, y_1, \text{string}, n, \text{size}, \text{frames}, \text{speed})$.

Rectangular-area Operations

paint The call-paint routine controls the painting operation in which every square in the programmed area is filled with a specified number representing one of the gray values or with the output from the random-number generator. The parameters for the routine are call paint (right, top, left, bottom, $\frac{R}{n}$).

shift The call-shift routine moves the entire contents of the rectangular area to a new location by specifying a number of squares in a stated direction. The parameters for the routine are call shift (right, top, left, bottom, $\frac{\text{right}}{\text{up}}\frac{}{\text{left}}\frac{}{\text{down}}$, amount).

tlit This computer-programming instruction is used for transliterating previously programmed numbers. The parameters for the routine are call tlit (right, top, left, bottom, table).

cømbin This programming instruction is used to order the combination of picture elements from two grid areas. The parameters are call cømbin (right, top, left, bottom, onto-right, onto-top, $\frac{\text{big}}{\text{small}}\frac{}{\text{sum}}\frac{}{\text{diff}}$).

center This rectangular-area routine controls the centering of objects, usually typed captions or labels, within a specified rectangular subarea within the grid. The parameters are call center (right, top, left, bottom).

cøpy This programmed command is used to copy an area, specified by the first four parameters, to a second location in which the top right-hand corner is defined by onto-right, onto-top. The parameters for the routine are call cøpy (right, top, left, bottom, onto-right, onto-top).

expand This rectangular-subarea routine enlarges or expands the previously programmed picture by the indicated factor. The expansion is centered at (x_c, y_c). The parameters are call expand (right, top, left, bottom, x_c, y_c, factor).

smøøth This rectangular-subarea instruction is used to smooth the picture by rounding sharp corners or filling in areas that need patching. The parameters for the routine are call smøøth (right, top, left, bottom).

fringe This programming instruction is used to draw a fringe or outline around objects. It is accomplished by programming the next higher number around an object or area that needs definition. The parameters for the routine are call fringe (right, top, left, bottom, next-to, becomes, flag).

fill This rectangular-area routine is used to fill or opaque a specific area with a gray tone. The parameters are call fill (right, top, left, bottom, x, y, n, boundary-n).

zøøm This programming instruction is used to produce an effect similar to that achieved with a zoom lens. The programmed image is dynamically expanded by the indicated factor, with the expansion centered at (x_c, y_c). The parameters for the routine are call zøøm (right, top, left, bottom, x_c, y_c, factor, frames, speed).

Miscellaneous Operations

camera This programming command controls the output operation. The parameters for the routine are call camera (frames). If the installation provides for various options, the parameters must be preset by other subroutines.

debug This programming command prints out a subarea and lists all of the bug positions within that area. The parameters are call debug (right, top, left, bottom, width, height).

update This programming routine updates and stores the previously programmed picture elements on the surface of the two-dimensional grid.

bgread This programming command updates the positions of all image-forming elements on the grid after a subroutine operation.

letter This routine, along with type, is used to control the formation of patterns and the display of alphameric characters by numerically encoding the string of entries representing each of the numbers and letters. The routine is expressed as letter (n, string).

randøm If restart equals zero, this operation represents the next output from the random generator. The routine is expressed as randøm (restart). Randøm and randwd both use a computer-stored comman table, ir, of randomly selected numbers. The random-number generator, randnø, assigns numerical equivalents representing starting points, ist; skip distances, ipd; and working numbers, nø. The ist and ipd numbers are indexed into the table ir to generate new starting points, skip distances, and numbers or parameters. If new ranges or number distributions are required, the entries in the table must be changed accordingly.

randwd The function of this routine, an acronym for random word, is to reset the random-number generator when restart equals zero. The routine is expressed as randwd (restart).

randnø The symbol for the random-number generator.

The higher-level commands for performing drafting-type operations and for manipulating the graphics display, along with subroutines for miscellaneous purposes, are listed in the following table.

grid operations	windøw	repositions the window represented by the internal grid
bug operations	place	places a bug at x, y coordinates
	møve	moves a bug
drafting operations	rect	draws a rectangle
	line	draws a straight line (dynamic)
	arc	draws an arc or circle (dynamic)
	trace	traces other curves (dynamic)
	type	types alphameric characters (dynamic)
rectangular-area operations	paint	uniformly paints an area
	shift	shifts contents of an area
	tlit	transliterates numbers
	cømbin	combines two areas
	center	centers objects (usually typed captions)
	cøpy	copies an area
	expand	enlarges a picture by an integer factor
	smøøth	smooths sharp corners
	fringe	puts a fringe around object(s)
	fill	fills a bounded area
	zøøm	approximates a zoom effect (dynamic)
miscellaneous operations	camera	the installation-supplied output routine
	debug	debugging aid—prints out a subarea and bug locations
	update	repacks words bugs are on
	bgread	unpacks words bugs are on (updates bugs after a subroutine)
	letter	function for getting n^{th} letter of a string
	randøm	function for delivering random numbers
	randwd	function for delivering a full word of random numbers

grid operations	windøw	subroutines and parameters (x-min, y-min)
bug operations	place	(bug, x, y)
	møve	(bug, direction, how far)
		tø, up, døwn, left, right, nørth, east, søuth, west
		(direction)
drafting operations (n = number)	rect	(right, top, left, bottom, n, width)
	line	$(x_1, y_1, x_2, y_2, n, width, frames, speed)$
	arc	$(x_1, y_1, x_c, y_c, sense, n, width, x-limit, y-limit, frames, speed$
		sense: cw, clockwise; ccw, counterclockwise
		width: 1, 2, 3, 4, 5, 6, . . .)
	trace	$(x_1, y_1, string, n, width, frames, speed)$
		Hollerith string or alphamerics)
	type	$(x_1, y_1, string, n, size, frames, speed)$
rectangular-area operations	paint	(right, top, left, bottom, n)
	shift	(right, top, left, bottom, direction, amount)
		up, døwn, left, right (direction)
	tlit	(right, top, left, bottom, table)
	cømbin	(right, top, left, bottom, onto-right, onto-top, rule
		rule: big, small, sum, diff)
	center	(right, top, left, bottom)
	cøpy	(right, top, left, bottom, onto-right, onto-top)
	expand	$(right, top, left, bottom, x_c, y_c, factor)$
	smøøth	(right, top, left, bottom)
	fringe	(right, top, left, bottom, n, next-to, becomes, flag)
	fill	(right, top, left, bottom, x, y, n, boundary-n)
	zøøm	$(right, top, left, bottom, x_c, y_c, factor, frames, speed)$
miscellaneous operations and functions	camera	(frames)
	debug	(right, top, left, bottom, width, height)
	update	(repacks words bugs are on)
	bgread	(unpacks words bugs are on, updates bugs after a subroutine)
	letter	(n, string)
	randøm	(restart, function for delivering random numbers)
	randwd	(restart, function for delivering a full word of random numbers)

GENIGRAPHICS

General Electric's **Genigraphics** is an image-generation system for the creation, manipulation, digital storage, review, and communication of full-color graphics. Perfected in 1972, the system is an outgrowth of GE's experimentation in the field of visual-simulation technology. The primary elements of the Genigraphics image-generation system are the artist's console, the graphics processor, and the associated programs and graphics data. The artist's console is the focal point of the system.

The console is designed and engineered to permit high-speed artwork creation, composition, and layout. Easy-to-use control devices and English-language communication enable the technique to be learned quickly and manipulated efficiently. The artist uses various keyboard and control-key functions to create images or to call up and modify prestored formats. Image components such as text, lines, areas, and data-plotting elements are entered via simple artist-oriented keyboard-input routines or by image descriptors prepared off-line on data cards or other input media.

Interactive control begins with cursor selection of the objects to be controlled. Visual-reference rectangles, indicating the precise size and position of the objects, provide the artist with a representation in real time of his artwork. Control-stick manipulation rapidly modifies the entire image or any portion. Operating modes and control knobs allow either a smooth transition through size, position, and color adjustments or a transfer of the characteristics of any one image component to other components. The color monitor permits constant inspection and appraisal of artwork as the artist progresses through the creative and interactive processes.

Artwork storage and retrieval, general inventory, and photography are also controlled from the console. Two magnetic-tape transports provide convenient high-density artwork storage and simple duplication and merging of images. An image recalled from temporary storage can serve as a fresh starting point for alternate aesthetic treatment, for integration with a graphic under development, or for restarting an alternative artistic exploration. Since the artwork is recorded in digital form, the artist is assured of consistent image quality during storage, retrieval, or transmission. With this system the artist can build his own library of design elements and finished art that can be instantly retrieved for any job. He can edit, replace, and modify these elements to achieve a wide variety of effects; he can zoom, crop, assemble, shade, isolate, reduce, and adjust the proportions of a single element or the entire design.

An added feature of the system, **Autocharts,** is designed to satisfy recurring graphics requirements and update specialized charts. The key to this system is a slide catalog of prepared chart outlines (graphic formats), through which designated colors and designs can be produced rapidly to fit individual needs. The prepared charts—format recommendations and color schemes—are illustrated in the catalog and number-coded for quick, easy reference.

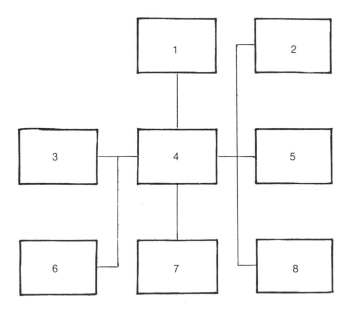

7-12. Genigraphics configuration: (1) storage-retrieval options, module identification (500), (2) photorecorder options (300), (3) artist's console (101), (4) graphic processor (201), (5) photorecorder (301), (6) input-device options (100), (7) programs and graphics (400, 700), (8) video recording-display options (600).

a

7-13. (a) The Genigraphics image-generating control console at Consolidated Film Industries, Hollywood, California. (b) Interactive controls. (c) Keyboard controls.

The system's 4,000-line high-resolution film recorder yields hard copy, film, or slides. The artwork can be transmitted over standard telephone lines or transferred as magnetic tape to a processing center for film recording. Optional equipment is available for television-compatible color-video transmissions, recording, and remote display of console-created or digitally stored artwork.

With the Genigraphics **interactive programs** the artist-operator creates image components at the keyboard through a series of simple question-answer routines presented on the keyboard display. These routines use the English language and are designed mnemonically to minimize memorization. The following is a list of Genigraphic's graphics-oriented initialization codes with their definitions and functions.

INTERACTIVE CONTROLS

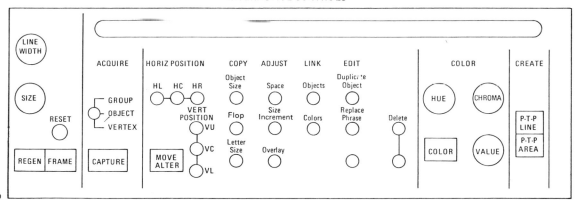

b

KEYBOARD CONTROLS

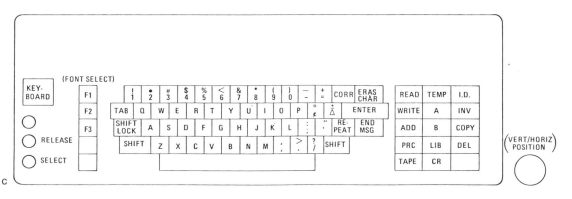

c

GT40

The **GT40** system, based upon the PDP 11/10 mini-computer and an interactive graphics scope, lends itself to real-time animation displayed on a CRT. Because of its small size programs tend to be written in assembly language. An interesting feature of the system is its graphical processor, which can access the computer's memory directly. It permits variations in line intensity, italicization, line blink, and the like without interfering with the operation of the standard central processor. A larger XVM medium-scale computer, used for interactive graphical design, can be adapted to the generation of animation artwork.

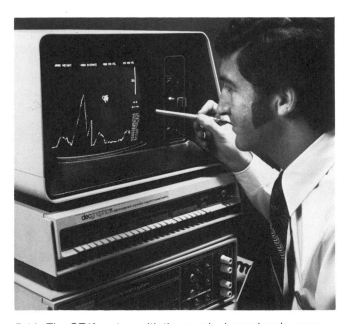

7-14. The GT40 system with the popular Lunar Lander game, which represents in pictorial animation a simulated Apollo landing, using actual NASA parameters. The action presented is performed in real time. (Courtesy of Digital Equipment Corporation.)

KEY-FRAME COMPUTER-GENERATED ANIMATION

Key-frame computer-generated animation techniques are analogous to conventional processes used by an animator to prepare key drawings in an action sequence. The animator's drawings (extremes) are used by his assistant (inbetweener) as a guide for the intermediate drawings needed to complete the action. Except for the hardware involved in electronic-imaging processes, the generation of key frames by computer follows the same general pattern. The system does not require the animator to become a computer programmer: in fact, any contact or communication between the animator and the computer-controlled system is through the artwork, which is displayed directly on the CRT (cathode-ray tube).

Although most computer-generated imaging techniques are language-driven, the key-frame system for animation sequences is picture-driven. The key drawings created by the animator from either the script or the storyboard are used to program the required action. The animator generally prepares the artwork for exposure at key intervals. During the playback the flow of action is previewed on the CRT display, and the intermediate positions computed by interpolating between the key frames are included in the display. Timing information indicating the number of frames needed to develop a smooth flow of movement is fed into the computer at the same time, as the key drawings are exposed. After the entire sequence of action is developed, it can be monitored or previewed at real-time rates. As in the conventional pencil test, changes may be made, reprogrammed, before the sequence is subsequently photographed on film or recorded on videotape.

The same techniques developed for conventional cel-animation production are the basis for key-frame animation. In place of the punched sheets of drawing paper mounted on the registration pegs

→

7-15. A reproduction of the computer-generated poster prepared for the 1973 Annecy Festival. (Courtesy of the National Research Council, Ottawa, Canada.)

9ᵉᵐᵉ JOURNÉES INTERNATIONALES DU CINÉMA D'ANIMATION

13-17 JUIN 1973

COMPOSITION DE PETER FOLDES SUR L'ORDINATEUR DU CONSEIL NATIONAL DE RECHERCHES DU CANADA

of the underlit animation board the animator's key drawings, or extreme positions, are displayed directly on the CRT. The type of computer-generated figure best suited for this technique is a line drawing. The figures produced by electronic means reflect the animator's style and talents just as the more conventional production processes do.

The lines on the image-forming key drawings may be subdivided into groups of lines on a number of cels if necessary in order to facilitate programming and storage. The component cels, each with its own group of lines, are programmed in their precise positions in relation to the other cels needed to form any of the key-frame composite images. Individual instructions regarding the required motion of each of the component cels are programmed separately. Similarly, specific inter-polation laws may be assigned to each of the cels making up the composite drawings. These interpolation laws can be compared with the guides on an animator's extreme drawing that indicate how the inbetween drawings are to be spaced in order to produce the timing and action specified by the animator. The inbetween frames computed by interpolating between key frames are displayed during the playback at the cine rate. The entire sequence of action displayed on the monitor can be previewed, and any necessary changes reprogrammed immediately.

The hardware for the process consists of a high-speed digital computer and the cathode-ray tube used to display the sequence of action. The memory capacity of the processor is adequate for the most complex types of animation. The software

7-16. Computer-generated animation from *La Faim*, a film by Peter Foldes. (Courtesy of the National Research Council, Ottawa, Canada.)

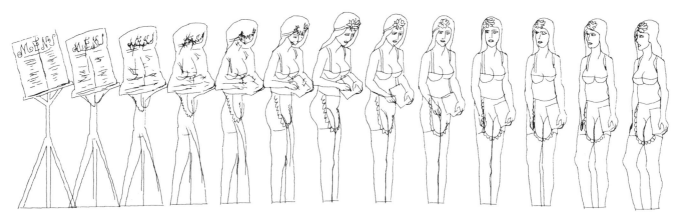

7-17. This sequence of images was extracted from *La Faim*. Produced by René Jodoin of the National Film Board, Canada, the interpolated images reflect the same artistic style as the extremes, or key frames, drawn by the animator. The film received the Prix du Jury at Cannes in 1974, the Golden Hugo in Chicago, and a special prize at the Barcelona Film Festival.

The film was also nominated for an Academy Award in the animated-shorts category. Unfortunately, the film's producers couldn't find a computer capable of making an acceptance speech! (Courtesy of the National Research Council, Ottawa, Canada.

consists of a number of interactive graphics programs that are used to create, modify, and manipulate the graphics fed into the computer.

Each of the lines in any of the component cels forming the composite key frame is made up of a number of small, interconnected lines. Each of these small lines, as seen in the reproduced display, is connected to x- and y-coordinates, coded with regard to intensity, and assigned separate interpolation laws that control its movement throughout the programmed sequence of action.

Communication between the animator or graphic artist and the computer is initiated with the artwork. Using industry terminology, the animator is able to manipulate the picture components and combine them with geometrical shapes generated by the computer. Preprogrammed routines and distortion functions are available in a **menu** from the computer's storage bank, which also includes a library of effects from previous productions. Access to such lists in this type of filing or retrieval system provides the animator with a great time- and budget-saver.

Since each of the lines in the interpolated images is produced in its entirety, lines that would normally be hidden behind some part of a solid, dimensional figure or object are clearly visible— one of the disadvantages of the technique. Similarly, lines that would be invisible behind an opaque or solid shape are clearly defined behind or through the interpolated transparent images. It is visually confusing, for example, to see part of the background through a figure. The obvious way to eliminate the undesirable see-through effect is to use some form of opaquing. Introducing or generating areas of color in specific locations is one way of solving the problem. Without departing from the basic technique of animating line drawings, conversion to a raster-scan format can be applied on a frame-by-frame basis to the computed image sequence. In this way any area that is completely bounded by an outline is filled in by adjacent horizontal lines. The filled-in areas obtained as a result of the scanning process serve as mattes and are used in the final display to eliminate transparency. An alternative method for removing transparency from computer-generated outline images is to use the low-resolution TV-data format for producing mattes. The random-line data is compared on a frame-by-frame basis with its corresponding blocking matte. Transparency is removed by controlling visibility with the low-resolution matte as each of the component cels is merged in turn to produce the composite image.

At the present time the cartoon characters used in the Alka-Seltzer commercials are excellent examples of the type of animation figures available with this technique. The computer is not yet a match for the artist capable of producing Disney-type animation.

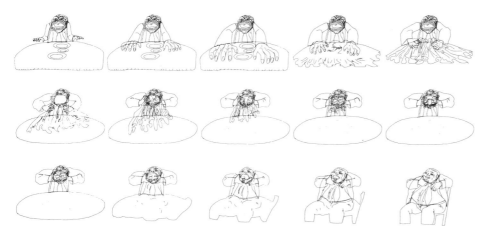

7-18. Key frames from an animated sequence in *La Faim*. (Courtesy of the National Research Council, Ottawa, Canada.)

OFF-LINE COMPUTERIZED ANIMATION RECORDING

The computer has eliminated practically all of the time-consuming tedium that characterizes the operation of the conventional animation stand. Before computers the animation cameraman, in addition to following cel-change instructions on the animator's sometimes indecipherable exposure sheets, plotted zoom and pan movements, fades, and dissolves (manually, of course) and consumed many cups of coffee in an attempt to meet impossible deadlines. Before the first frame of film could be exposed, the hapless cameraman checked and rechecked start and stop positions, took the veeder-counter readings needed for calibrating stop-motion zooms and pans, and racked the camera over to its nonfilming position for a final look through the viewfinder. The ever-present possibility of human error and the resulting inevitable retake hung over the cameraman's head like Damocles' sword. Computer electronics has removed the drudgery from animation-filming procedures and restored the smile to the cameraman's face. Through the magic of this modern miracle the animator or animation cameraman types filming instructions, using industry terminology, to the computer via a teletype—and all systems are go.

The computerized animation-recording system consists of a teletype, an automatic tape punch, a digital computer with a vocabulary of approximately 8,000 words, a camera control console, and

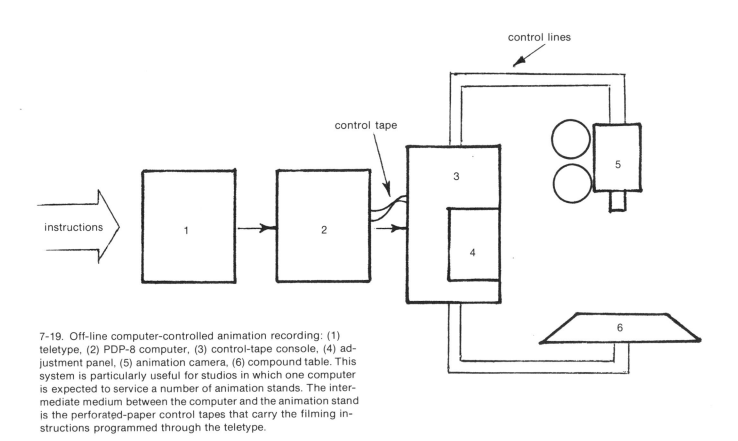

7-19. Off-line computer-controlled animation recording: (1) teletype, (2) PDP-8 computer, (3) control-tape console, (4) adjustment panel, (5) animation camera, (6) compound table. This system is particularly useful for studios in which one computer is expected to service a number of animation stands. The intermediate medium between the computer and the animation stand is the perforated-paper control tapes that carry the filming instructions programmed through the teletype.

the animation stand itself. The filming process actually begins when the computer's memory bank translates the teletyped instructions into **binary-coded decimals** (BCD). A high-speed tape perforator produces the tape, with the BCD numerical equivalents, that is fed into the console that controls the animation stand's camera and compound-table movements and functions. In addition to providing precisely repeatable incremental movements as small as $1/1000''$ the computer-controlled animation stand has an autoamtic lens iris, an automatic platen lift, and specially built pulse-action motors that control all north/south, east/west, and zoom movements.

The lengthy exposure sheets that contain the animator's filming instructions are no longer necessary. Plotting times for programming complex camera moves are greatly reduced, and dry runs for previewing movement are accomplished quickly and easily. If changes or revisions are needed, only those sequences of action that are involved have to be reprogrammed.

In **indirect,** or **off-line, computerized animation recording** the tapes are fed to a control console, which is connected to the animation stand. Since the computer is not connected directly to the animation stand, it is free to produce the perforated-paper control tapes for other filming projects—an obvious advantage and an important consideration for studios in which several animation stands are in constant use. With this system for recording animation the animator is able to plan complicated sequences of action that would be impossible or highly impractical without the precise control provided by the computer. Exceedingly troublesome moves, such as superimpositions combined with multiple runs or zoom and pan movements programmed over a skip-frame action, can be achieved easily.

Built-in warming systems ensure error-free operation. Even the programmer's instructions are checked as the information is teletyped, and prompt warnings urging the animator to review his instructions are not uncommon. Checks against panning rates that would produce a strobe effect, for example, are built in: the computer sends back a request to change speed and reprogram. Similarly, zoom movements that extend beyond the designated length of a scene are rejected, and the programmer is asked to review his instructions. The system also warns the cameraman if the shutter is closed or if the camera is set for a reverse rather than a forward mode.

The artwork mounted on the registration pegs in the animation stand's compound table may be given a dry run in order to check the flow of action. For this type of preview the positioning controls are reset to their start positions, the camera is racked over to the nonfilming mode, and the movement is monitored through the viewfinder. If the programmed action is satisfactory, the camera is returned to its filming position, the control tape is activated, and filming begins.

Since the filming process is no longer a tedious, time-consuming race against a deadline, both the

CLIENT: CBS SHOW "LOOK UP AND LIVE"
SUBJECT: "POETRY AND PHOTOGRAPHY"
PRODUCER: CHALMERS DALE
CONCEIVED BY: DOUGLAS STERLING PADDOCK
ANIMATION AND PROGRAMMING: AL STAHL/ANIMATED PRODUCTIONS, INC.

COMPUTER FOTOMATION EXPOSURE SHEET

PAGE 41 "LIGHTHOUSE SEQUENCE"

SCENES	ACTION	FRAMES
1	OPEN SHOT BEACH HOLD	0000
	START ZOOM PAN NORTH	0036
1A	START DISSOLVE	0125
2	CUT TO CU OF WAVE	0160
3	CHANGE PHOTO	
	CYCLE 2&3 FOR ANIMATION	0300
4	CUT TO DIAGONAL PAN	0420
4A	END PAN DISSOLVE TO	0640
5	LS LIGHTHOUSE/SLOW ZOOM IN	0740
6	CUT TO BIRDS	0800
7	CYCLE 6&7 BIRD PHOTOS	0940
8	CUT MCU LIGHTHOUSE	1000
8A	TRUCK BACK/FULL FIELD	1060
9	CUT TO DIAGONAL PAN	1100
9A	END PAN AND DISSOLVE TO	1140
10	VERTICAL PAN NORTH DOORWAY	1210
10A	END PAN	1240
	HOLD	1300
11	CUT TO CU LIGHTHOUSE	1360
12	MATCH CUT TO LIGHT ON	1400
	FADE OUT	1440

CHANGE TAPE TO SEQUENCE 42 "LIGHTHOUSE SEQUENCE" PART 2

7-20. Control tape for an animation stand.

animator and the animation cameraman are free to combine their skills to produce a more imaginative and creative film. At any time during the actual filming process the animator can make revisions to improve the total effect by reprogramming portions of the film. This capability is extremely important if a client is standing behind the cameraman and second-guessing every phase of the process. The animator or animation cameraman takes the last-minute suggestions in stride, reprograms that part of the sequence to satisfy the client, and filming continues.

7-21. Model 5442-C animation stand. (Courtesy of Oxberry— division of Richmark Camera Service.) This stand is an on-line computer-controlled animation system. It permits direct, precise computer control over the entire operation of the stand. The operator can direct the computer to execute complex sequences of animation photography with simple, compact instructions in animation photography with simple compact instructions in animation language.

Motion-picture and television production, whether a 2-hour feature film or a 1-minute commercial, generally begins with a concept and (subsequently) an approved script. The computer-controlled filming process follows the same pattern. In addition to the script or storyboard animation production is preceded by the soundtrack recording and its analysis by the film editor. Using a sound reader and synchronizer, the editor literally measures the prerecorded sound track that will accompany and complement the animation. The resulting frame counts are used by the animator in preparing the bar, or lead, sheets that account for every frame of film in a scene. These sheets are basically a synopsis that indicates where specific actions take place within a scene, the number of frames allotted for the particular action, the timing for synchronizing a cartoon character's mouth movements or lip sync with the prerecorded sound track, the frames in which camera effects such as zooms and/or pans occur, and the frame in which the scene cuts, fades, or dissolves into the next.

In order to increase the efficiency of computer-controlled filming, the sound track can be mixed on 1/4″ tape instead of magnetic film, thereby eliminating transfer costs. Since each scene in the animated production is usually photographed in sequence using exact frame counts, the entire editing process is greatly simplified. Splices are few and far between, and in many cases work prints in any form are completely unnecessary. The control tape is threaded into the control console, and the numerical equivalents are decoded by a tape-reading device. The digital-coded instructions (the frame counts provided by the film editor) are converted into electrical signals, which motivate the mechanisms that regulate the animation stand's camera and compound-table movements and functions—the north/south, east/west, and zoom positioning motors.

In practice, the animator or computer programmer and the animation cameraman discuss the scene's requirements. The general flow of action in relation to the static artwork is planned in terms of camera movement, and the instructions are fed to the computer via the teletype. Coordinates needed for plotting camera movement are obtained by sighting through the viewfinder and noting the corresponding numerical positions of the north/south, east/west, and zoom veeder counters which operate in conjunction with the respective positioning motors. After the coordinates for both the start and stop positions are determined, the computer makes the necessary calculations, based on the input information, and plots the movement from one position to another. The number of frames allotted for the movement is used as the guide for determining the speed and spacing between each of the individual stop-motion moves. Abrupt starts and stops at the beginning and end of zoom and pan movements are eliminated by programming acceleration and deceleration moves for the first and last frames of the effect. The ease-in and ease-out add a professional touch to any zoom or pan movement.

This type of programming eliminates the need to plot the positions and moves for each piece of artwork in a sequence in turn. With computer techniques as many as 50 separate pieces of artwork can be arranged as a unit. Any type of graphic information or a combination of artwork and overlay cels containing lip-sync or cyclic action can be included at any point in the display. During the filming process the camera is guided from one piece of artwork to the next by the punched control tape. The positioning motors locate each of the preprogrammed field positions quickly and precisely. This type of error-free filming can be accomplished in a fraction of the time required for conventional methods and equipment.

The camera shutter, also activated by the control tape, is used to produce precisely metered fades and dissolves in any specified number of frames. Information fed to the computer with regard to any film's characteristic curve allows it to make the necessary compensations that assure constant, total screen brightness even if the dissolving scenes range from one extreme to another during the transition. Similarly, a fade-out can be programmed to eliminate the rapid drop-off that usually occurs over the last few frames of the effect.

Object-creation Routine Initialization Codes

feature	code	definition	function/input
artwork set-up	clr	color	color-scheme selection
	bac	background	background (artwork list cleared)
	frm	frame	film-format-frame select
text	txt	text	operator-specified size, position, color of text
	dtx	default text	preprogrammed size, position, color of text
	ttl	default title	preprogrammed size, position, color of title
	tab	tabular	facilitates input of columnized data to operator-determined defaulted position, size, color
shapes	rec	rectangles	operator-specified size, position, color
	cir	circles	operator-specified size, position, color
	sec	circle sectors	operator-specified size, position, color, orientation
	are	area	operator-specified vertex description of geometric area
	lin	line	operator-specified vertex description of nongeometric line

Data-plotting Routine Initialization Codes

	code	definition	function/input
data plot areas	cpl	circular plot area	preprogrammed plot area for pie plots
	rpl	rectangular plot area	preprogrammed plot area for various plot types such as line, area, bar plots
rectangular plot area setup	vsv	vertical scale values	automatic placement of operator-specified ordinate for vertical data plot
	hsl	horizontal scale labels	automatic placement of operator-specified abscissa for vertical data plot
	vsl	vertical scale labels	automatic placement of operator-specified ordinate for horizontal data plot
	hsv	horizontal scale values	automatic placement of operator-specified abscissa for horizontal data plot
rectangular plot area enhancement	vst	vertical scale title	ordinate title in preprogrammed size, position, color
	hst	horizontal scale title	abscissa title in preprogrammed size, position, color
	vsg	vertical scale grids	automatic placement of horizontal grids or ticks derived from vertical scale values or labels
	hsg	horizontal scale grids	automatic placement of vertical grids or ticks derived from horizontal scale values or labels
	vsd	vertical scale divisions	automatic colored horizontal subdivision of rectangular plot area as determined by number of values or labels in vertical scale
	hsd	horizontal scale divisions	automatic colored vertical subdivision of rectangular plot area as determined by number of values or labels in horizontal scale
rectangular plot area types	vbr	vertical bar plot*	all plot types are automatically positioned in the rectangular plotting area to the parameters established by the vertical and horizontal scale inputs. The lateral positioning of bar plots and data point locations of line and area plots is under operator control.
	hbr	horizontal bar plot*	
	arp	area plot*	
	lnp	line plot	
	lua	lower/upper area limit plot	

*cumulative value capability

116

Operational Modes and Controls

Console capabilities are controlled from the keyboard with a number of interactive controls. These controls are arranged into seven operational modes: capture, move-alter, regeneration, color, frame, point-to-point area and line, and keyboard. The individual controls, arranged by modes, are described below.

†LEGEND 1 bidirectional x/y rate control sticks spring loaded to zero
2 bidirectional rate control knobs spring loaded to zero
3 pushbuttons
4 three-position toggle switch
5 function keys

capture mode—provides the artist with cursor-selected control of any objects in the artwork

Control	type+	function
right control stick	1	x/y cursor positioning for object selection
select	3	operator-to-processor signal for control of cursor-selected object
release	3	operator-to-processor signal to release control of last selected object
group/object/vertex	4	provides operator access with single and multiple object or vertex control

move-alter mode—places captured objects under operator control via rate controls or function pushbuttons

horizontal position		
horizontal left (hl)	3	left alignment to first object captured
horizontal center (hc)	3	horizontal center alignment to first object captured
horizontal right (hr)	3	right alignment to first object captured
vertical position		
vertical upper (vu)	3	top alignment to first object captured
vertical center (vc)	3	vertical center alignment to first object captured
vertical lower (vl)	3	lower alignment to first object captured
copy		
object size	3	vertical and horizontal dimensional extremes copied from first object captured
letter size (text only)	3	letter size and aspect copied from first object captured
Adjust		
space	3	position of objects evenly spaced between first and last objects captured
size increment	3	size and aspect of objects incremented between first and last objects captured
overlay	3	adjustment of visual priority between any two objects captured

117

link		
objects	3	links two or more objects for convenient size or position control acquisition with single select signal
colors	3	links color of two or more objects for convenient color control acquisition with signal select signal

edit		
duplicate	3	duplication in the image of one or more objects captured
delete	3	deletion from the image of one or more objects captured
flop	3	left/right orientation reversal of one or more objects captured (text not included)
replace	3	copyediting or replacement that preserves size, position, and color of text line captured
left control stick	1	independent x and y control of object or group size
right control stick	1	independent x and y control of object or group position
size	2	proportion-maintained x/y control of object or group size
line width	2	control of line width

regen mode—used to regenerate or refresh the image subsequent to operator adjustments and/or data entry

color mode—gives the artist color control over selected objects through independent hue, chroma, and value controls

control	type+	function
right control stick	1	x/y cursor positioning for object selection
select	3	operator-to-processor signal for color control of cursor-selected object
release	3	operator-to-processor signal to release control of and restore previous color to object selected
hue	2	continuous smooth hue adjustment of object(s) color
chroma	2	continuous smooth chroma adjustment of object(s) color
value	2	continuous smooth value adjustment of object(s) color

frame mode—presents a preselected film-format outline on the display for image composition: size and position is under operator control for zoom, crop, and overall image distortion

left control stick	1	independent x or y size control of frame
right control stick	1	independent x or y position control of frame
size knob	2	proportion-maintained x/y size control of frame
reset	3	automatic restoration of original frame to artwork/position relationship

point-to-point area and line mode—used for convenient interactive creation of areas and lines using cursor-established vertex locations

right control stick	1	x/y cursor positioning for vertex or line location
select	3	operator-to-processor signal for vertex or line creation
release	3	operator-to-processor signal for release of last vertex or line selection

keyboard mode—keyboard creation of image components (in directing data input), font selection, and artwork-record management

key groups	type†	function
keyboard		standard typewriter keyboard with a complete set of alphamerics (upper- and lower-case letters, numbers, and special characters); special keys for individual character of full line editing
font select keys	5	rapid selection of or changeover to either of three preselected software-controlled font styles
transfer functions	5	read—artwork retrieval for display on console monitor write—storage of displayed artwork add—artwork retrieval for addition to image displayed on monitor prc—automatic film recording of artwork
peripheral selection	5	temp—temporary artwork storage area a—tape deck a b—tape deck b lib—library of artwork on disc file cr—card reader
file function	5	id—identify source by tape/tape deck designation inv—keyboard display of tape directory copy—hard-copy printout of tape directory del—deletes artwork records from tape

The shutter can also be used to eliminate one of the most monotonous, time-consuming assignments in animation photography—the filming of cyclic action. The constant repetition involved in filming even the simplest cyclic action is tedious and fatiguing, and costly retakes are quite common. An 8-frame animation cycle (the first frame is drawn so that it hooks up with or follows the eighth frame in a continuing action), repeated for a 3-second sequence, would normally involve a minimum of 72 cel changes—and background moves if a panning action is required. Computer technology virtually eliminates this frustrating process by providing a skip-frame capability through controlled shutter movement. During the filming process the first cel in the repeat cycle is positioned on the compound table's registration pegs over the scene's background. The control tape is programmed to expose cel 1 at frames 1, 9, 17, 25, 33, 41, 49, 57, and 65. The shutter closes automatically after each of these frames is photographed, and the film is advanced for the next exposure. The shutter is closed after cel 1 is exposed for the last time at frame 65, and the film is returned to frame 2. Cel 1 is replaced by cel 2 on the registration pegs, and the skip-frame process is repeated, with cel 2 photographed on frames 2, 10, 18, 26, 34, 42, 50, 58, and 66. The shutter is again closed, and the film is returned to frame 3 to film the cycle with cel 3. The process is repeated until each of the cels in the cycle is photographed at the appropriate frames. Filming the entire action for all 72 frames requires only 8 cel changes. If the background is panned during the cycle, the platen is automatically lifted between exposures, and the background advanced to its next programmed position. The entire filming process is completed in a fraction of the time that would be needed for conventional procedures.

The filming of cartoon animation, static artwork, photographs, and even three-dimensional objects for television commercials has always been considered routine for animation-stand photography. With the development of new techniques— the filmograph (limited animation) and the visual squeeze, for example—the use of 35mm transparencies increased greatly, but even the better-equipped animation stands were unable to photograph an area larger than 1″ field. Within the last decade, however, new lens systems have been developed that are not only capable of photographing a slide but can zoom into an area within the slide. Attached to the camera carriage, newly installed Nikkor macrolenses that follow focus automatically during zoom movements are able to achieve an 8:1 zoom ratio. A 1:1 zoom ratio indicates the exposure of an area equivalent to the gauge film stock used for shooting. Since the subject matter requires proportionately more light as the lens is moved closer, the animation camera was redesigned so that the aperture opens automatically to compensate for the closer distances between the macrolenses and the transparencies.

Indirect computer-controlled animation-recording systems have a number of advantages compared to both conventional and direct, on-line systems. In addition to their speed and error-free operation the programmed instructions are precisely repeatable: the control tape can be filed under its proper classification and retrieved for use at a later date. Similarly, complicated motion sequences may be cataloged and cross-filed to build a library of effects that can be used for other productions. With off-line systems tapes can be prepared while the animation stand is in use for other productions, a distinct advantage for busy studios in which a number of stands are in constant operation. The animator can discuss a scene with a computer programmer or animation cameraman and return to his underlit drawing board to plan his next move. Chances are better than good that the scene or sequence will be photographed before he finishes sharpening his pencil. Compare on-line computerized animation recording.

ON-LINE COMPUTERIZED ANIMATION RECORDING

On-line or **direct, computer-controlled animation recording** is basically the same as the off-line system. The main difference between the two lies in the relation of the computer to the animation stand. Unlike the off-line system, in which the control tape programmed by the computer is fed to a control console that is connected to the animation stand, the computer in the on-line system is connected directly to the animation stand via the control interface. This direct communication eliminates the need for an intermediate medium such as the punched control tape. Although the direct communication between the computer and the animation stand can be an advantage during the filming of long, complex sequences, the computer is tied up until the entire recording process is completed. In studies in which the computer is kept busy programming control tapes for a number of animation stands a great deal of waiting time may be involved. Compare off-line computerized animation recording.

SCANIMATE

Simply defined, a **video synthesizer** is a computer system that can convert all of the component parts of an image into electronic signals, modify the signals in compliance with the programmer's instructions, and display the continually changing image on a monitor. This is what the **Scanimate** system does. The Scanimate system is designed to fill the graphic artist's every need. The subject matter for the Scanimate pictorial presentation may include static artwork, abstract forms, photographs, titles, logos, or three-dimensional objects in both black and white and glorious living color. The specially designed electronic cameras used in this

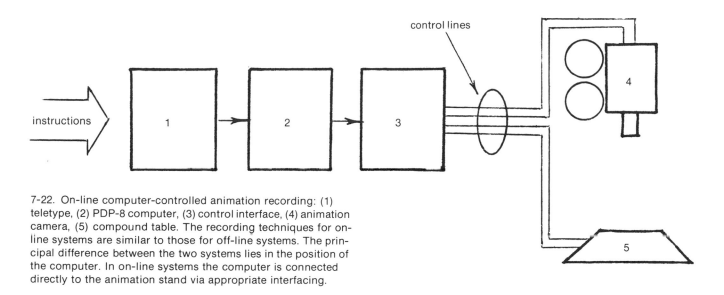

7-22. On-line computer-controlled animation recording: (1) teletype, (2) PDP-8 computer, (3) control interface, (4) animation camera, (5) compound table. The recording techniques for on-line systems are similar to those for off-line systems. The principal difference between the two systems lies in the position of the computer. In on-line systems the computer is connected directly to the animation stand via appropriate interfacing.

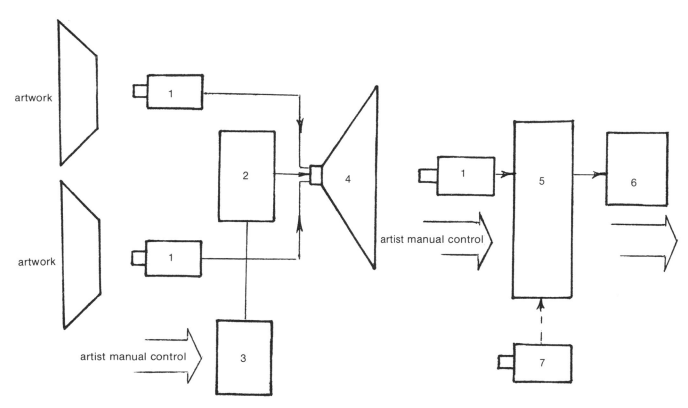

7-23. Scanimate computer system: (1) TV cameras, (2) raster generator, (3) computer controller, (4) high-resolution (x, y) display, (5) NTSC color, (6) color-TV display, (7) studio color camera(s).

system widen this graphic range still further by allowing the artist or director to introduce Kodaliths, slides, one or more frames from a reel of motion-picture film, or previously recorded videotape images. Disney-type animation is best left to more conventional animation-production techniques, but animation effects such as those used in TV science-fiction series—*Star Trek*, for example—or commercials—Alka-Seltzer, for example—are easily achieved with this computer-generated animation technique.

In most cases the artwork is prepared as a **Kodalith**—a black-and-white high-contrast transparency. Each component part of the one or more Kodaliths making up the composite image is recorded as a separate element and manipulated in relation to the movements programmed for the other component parts. The combined information is logged into the computer's memory bank. The interpolated moves plotted by the computer are inserted between the initial reference input (the original image exposed to the TV camera) and the stop position. The completely coordinated, fluid sequence of action is immediately available for display, preview, or recording either on film or videotape. Images fed into the computer's memory bank can be shrunk, stretched, squeezed, twisted, bent spatially, revolved, rotated, exploded, splintered, zoomed up to fill the frame, or zoomed back to infinity. A single image can be repeated as many times as the programmer wishes until it is duplicated throughout every part of the frame. Sharp corners can be rounded out by turning a dial, or razor edges put on round corners. Negatives can be combined with positives to create an unlimited variety of unusual illusory effects.

With the Scanimate system one illustration can be superimposed over another, or, alternatively, components can be subtracted and areas keyed out so that other images can be exposed in the matted areas. Color is handled in almost the same way as an artist might apply oil paints to canvas. Colors, if they need to be enhanced in any way, can be changed easily by the encoder that provides the controls for selecting and regulating chrominance and luminance. Blues can be turned into reds, and purples to oranges with little effort.

Solarization and other optical effects are also available by turning a knob, pushing a button, or flipping a switch. The visual effects and movements created in the computer can be played back and monitored immediately; changes can be made almost as quickly. The single most important ingredient in producing animated films and videotapes with Scanimate is imagination. The animator or programmer can produce an infinite variety of spectacular visual effects and put an image

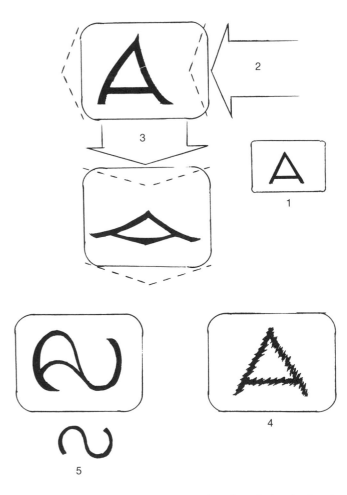

7-24. In the Scanimate system a raster-bending electronic signal causes the displayed image to follow the wave-form shape: (1) normal image, (2) delta wave, (3) delta wave, (4) high-frequency delta wave, (5) sine wave.

through an incredible range of movements by simply manipulating knobs and dials. Although it is basically a computer, Scanimate is primarily a creative tool designed to enable artists, not engineers, to animate images electronically.

The first step in the actual production process begins with a Kodalith or with an animator's previously prepared graphics. In either case the subject matter is exposed to the special TV camera, and the input is converted into electrical signals as in conventional electronic photography. The image is immediately visible on the studio monitor. Processed by the computer, the graphic artist is assured that the style and content of his carefully prepared artwork will be reproduced faithfully down to the most minute detail. For artists whose

work must be handled by other artists and technicians before it is finally reproduced, the very thought of losses in quality can produce nightmares. With this system, as Flip Wilson might say, "What you see is what you get."

The animator or programmer watches the image on the monitor screen as he activates it by manipulating appropriate controls on the animation control console. Each knob or dial affects a specific characteristic of the image such as position, size, intensity, duration of the animation sequence, and the animation itself. The artist usually controls the console adjustments and programs the movement rather than the studio technician. Using the graphics displayed to the TV camera as the reference input, the programmer first adjusts the in-

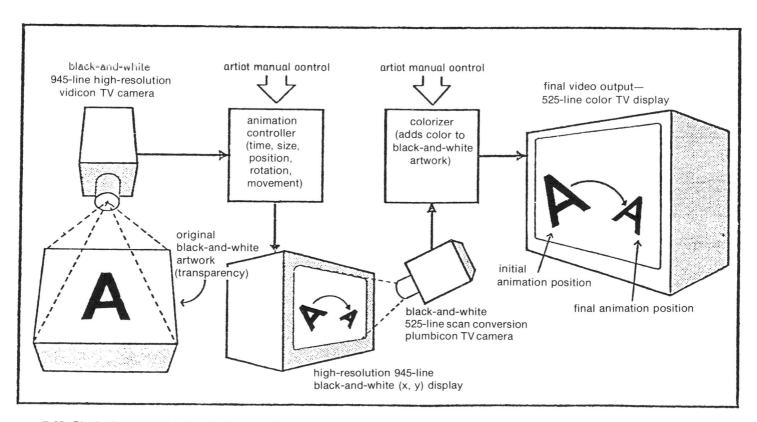

7-25. Block diagram of the Scanimate animation computer.
(Courtesy of computer Image Corporation.)

tensity and size of the image. With the **final** position controls the animator moves the image to the position on the screen at which he wants it to stop at the end of the animation sequence. After making similar adjustments with the **initial** position controls for the beginning of the action, the animator sets the length of the sequence to any interval from 1/2 second to several minutes. When the animation sequence is worked out and all of the adjustments are set, the image is returned to its starting position, and the **animation** switch is flipped. This triggers the animation cycle, and the image moves from the initial to the final position, simultaneously executing whatever animation is programmed. The sequence may be repeated as often as desired or changed as many times as is necessary until a satisfactory sequence is obtained.

It is assumed above that the entire image is animated as a whole. In many applications this is desirable, and the results are completely satis-

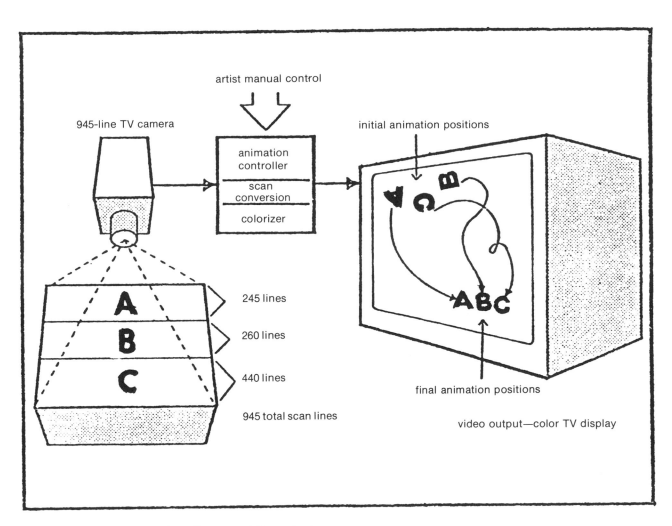

7-26. Multiple-image animation. (Courtesy of Computer Image Corporation.)

factory. But in an automobile engine, for example, pistons, valves, cams, and other parts have their own distinctive movements and timing. With the Scanimate system the whole may be divided into its component parts, and each part animated independently of the others. Each of the moves making up the composite is displayed in real time on the studio monitor. The entire production is built up sequence by sequence until both the animation and the color are judged to be optimal.

Operating at a 48-frames-per-second rate, the video output displayed on the CRT can be filmed with standard motion-picture equipment operating at sound speed, 24 frames per second. The motion-picture camera and the TV camera must be synchronized for the filming process. Alternatively, the computer can be operated at a 60-frames-per-second rate so that the CRT display may be viewed by a second TV camera and recorded on videotape. Compare Caesar.

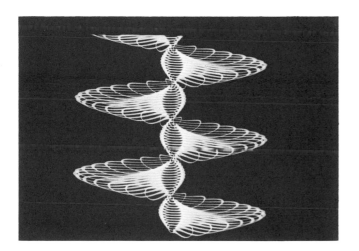

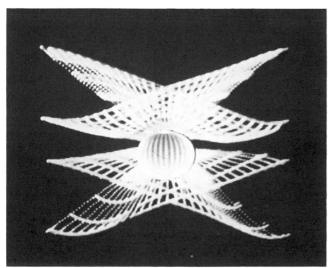

7-27. *From this to this*, computer generated animation effects. (Courtesy of Dolphin Productions.)

7-28. Computerized show openers. (Courtesy of Dolphin Productions.)

STATIC COMPUTER-GENERATED ANIMATION IMAGING

In **static computer-generated animation imaging,** also called static computer imaging and static display images, the computer is used to create images as well as to record them. Each coordinate of the CRT (cathode-ray tube) face on which the image is generated and displayed is represented by a numerical equivalent. Each of the x- and y-axis coordinates is capable of producing a dot of light. A number of dots, programmed in specific positions, can be joined to form a straight line, a perfect circle, or any combination of the two. Images are generated by assigning specific numerical coordinates to different points on the CRT display, with each bit of picture information precisely positioned by mathematical definition.

Two types of CRT are used for generating the static images—refresh scopes and storage scopes. The **refresh scope** must continuously retrace, or refresh, the generated image for it to remain visible. The process is similar to the continuous field renewal of television receivers. In the **storage scope,** by comparison, the electron beam in the CRT activates a phosphor that is capable of glowing for many minutes at a time. The storage scope is preferable for the formation of complex images. Lines that would normally be hidden behind a portion of a figure or solid object can be eliminated, along with the visual confusion created by their presence, during the programming process. The general technique permits sequential placement of many figures on the scope face, and each image can be photographed as it is presented on the CRT display. A solenoid is used to synchronize

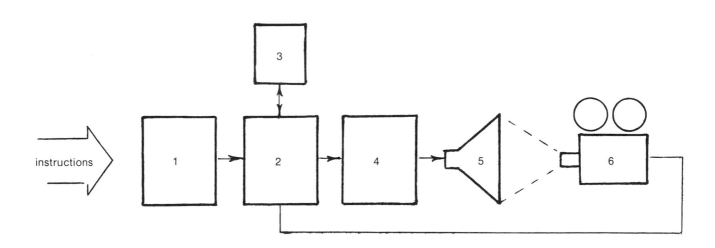

7-29. Static computer-generated animation recording: (1) teletype, (2) medium-to-large computer, (3) bulk memory storage, (4) display console, (5) display tube, (6) camera with incremental motor. In this system the computer creates as well as records the artwork. Mathematically defined image-forming elements are assigned specific coordinates and displayed on the CRT face. They can be moved incrementally to other assigned positions on the scope face in compliance with the programmer's instructions.

the camera shutter in the stop-motion setup with the computer's output during the photographic process.

Because the image-generating technique is based on mathematics the system is positive and precise. Every component part of any image created for a sequence of action can be displayed repeatedly, and, since the graphics are made up of numerical entries, they are for all intents and purposes practically indestructible. The obvious advantage of any system of this kind is that the image can be displayed while it is being generated. The disadvantage of any mathematical system, however, is the ever-present possibility of human error in programming or in writing equations for complicated sequences. Complex figures necessitate proportionately long plotting and calculating times and in many instances exposure-time requirements that create problems for conventional camera systems. Simpler figure formation and the simulation or display of images or moving bodies for scientific purposes are particularly well suited to this basic technique. Compare continuously variable display images.

SYNTHAVISION

The **SynthaVision** computer system for generating and animating shapes offers the programmer or graphic artist a sharper pencil along with a greater outlet for his imagination and creative talents. This technique translates simulated objects and figures into mathematical formulas that can be displayed as images on a cathode-ray tube for reproduction either by conventional motion-picture cameras or directly on videotape.

Developed by Dr. Phillip Mittelman, this sophisticated computer system is sometimes referred to as MAGI, an acronym for the Mathematical Applications Group, Inc. Originally used to provide realistic simulations for the atomic-energy-research program, the SynthaVision technique translates basic geometrical shapes into real or imaginary objects. The system can also convert abstract designs into moving models that show changes in perspective and three-dimensional effects.

The process of generating graphics begins with a series of IBM cards that carry entries describing specific geometrical shapes, the desired nonexistent lighting setup, and the required camera angle, which can be achieved with any combination of imaginary focal-length lenses. The generated image is formed bit by bit, with each component part assigned to a specific part of the frame. Also included on the IBM cards is information regarding the initial screen position of the composite shape and its subsequent movement along the x-, y-, and z-axes to its final position in the frame. Add the further possibilities of zoom, tilt, pan movement, rotation, and revolution—and the simulation of the concept or design assumes amazingly realistic proportions. Specific colors for the total design or for each of the component parts are also entered on the cards fed into the 360/65 IBM digital computer.

Animating titles and graphic designs generated with the SynthaVision technique can be splintered and just as quickly reassembled in response to the director's or programmer's slightest whim. Objects can be instructed to change sizes and colors as well. The input information is also used to determine the shape of the generated image from any angle so that it can be made to turn and change perspective in order to create a feeling of depth. Every component part of the image can be programmed to move independently in any direction, at any speed, and along any axes. Rotating and/or revolving the geometrical forms of the composite image produce changes in viewing angles. The computer "camera" can even be placed within objects to provide additional illusory effects.

Objects that have no counterpart in real life and newly designed cartoon characters that exist only in fantasy worlds can be created simply by describing them on the IBM cards. Similarly, prototypes of machines still at the drawing-board stage can be programmed to operate as they would be expected to when finally developed. The programmer describes the machinery, real or imaginary, and programs its operation. Since the generated simulation exists only in the programmer's imagination, it can be asked to do anything—even the impossible.

In operation the storyboard or script is analyzed

with respect to the details of each object, color, light source, and camera angle. The information is transferred to IBM punch cards and translated into basic geometrical shapes that are assigned to specific areas of the display. A computer picture tape, the result of extensive calculations based on the programmed information and instructions, is read into a minicomputer that actually generates the picture. The minicomputer also runs the high-resolution 1,000-line CRT on which the pictures are formed. Representative frames in the sequence may be seen on the monitor and previewed before the sequence is finally recorded on either film or videotape. Each picture in the sequence is exposed on a frame-by-frame basis through three primary-color filters in the computer-controlled color wheel. All the colors and subtle shadings are reproduced as faithfully as they would be with color-separation positives and conventional optical-printing processes. If live-action footage is to be combined with the images generated by the computer, a traveling matte of the action is produced with conventional processes for combination on the optical printer. The exposed film is developed normally by a professional motion-picture laboratory using standard procedures. If the dailies are approved, the IBM cards are filed so that a new original may be produced at a later date.

This relatively new technique has already established itself as an extremely useful tool in many fields and for many purposes. For the advertising agency engaged in the production of television commercials, for example, it has provided a new look. The art director can ask the computer to visualize a new product; make exciting new graphic presentations; animate logos, titles, and station call letters; and move products up, through, and around any part of the frame to create an infinite variety of visual effects. The effects made possible with this image-generating technique are limited only by the collective imaginations of the men and women who sit in director's chairs. Industry has also benefited greatly by using the technique to show how machinery works without stopping plant operations. Films for training and instructing personnel can be made without costly dismantling of equipment or interruption of production lines. Traf-fic patterns relative to the placement of new equipment before it is actually installed can be studied in order to eliminate the needless cost and inconvenience of relocation at a later date. Similarly, research and planning studies can save the taxpayer substantial amounts of money. Government agencies can use the technique to train personnel, simulate conditions associated with space exploration, and eliminate the costly process of building new armaments and sophisticated missile systems by generating a simulated version, testing it under different conditions, and evaluating performance before investing a single penny for a prototype. In engineering, architecture, and transportation the technique is used to simulate bridges, tunnels, and buildings. The Urban-renewal projects or the construction of an entire city can be planned, and planes, trains, and buses can be moved into and through an area before the blueprints are off the drawing board. In education the creation and manipulation of objects in single-concept film loops, slides, and filmstrips make the learning process a more pleasurable experience. In the art and entertainment fields, new forms for all age groups are being produced daily.

The developer of SynthaVision predicts a spectacular future for the new technique. The growing sophistication of computers is actually reducing costs in the audiovisual field. In the near future it is expected that the SynthaVision technique will be used to produce lifelike scenes on stereo slides as well as to record a picture on videotape or film. SynthaVision computers will be able to absorb photographic information for combination with real backgrounds and real people to produce dimensional realistic and surrealistic images.

TARPS

A computer language developed for generating graphics, **Tarps** is an acronym for two-dimensional alphameric raster picture system. The language basically consists of a set of macros written in terms of Beflix. Image-forming textures and tonal values are produced from arrays of closely spaced characters, which are identified by a numerical code.

TELEMATION

Telemation is an electronic titling generator that prepares printed graphic material almost instantaneously. Since the titling generator itself is the video source, no TV camera is required to pick up the prepared information.

Designed by Telemation, Inc., of Salt Lake City, the titling generator's characters may be mixed, supered, keyed, or wiped in the same manner as any other video source. In this way the titling-generator output can supplant the use of superimposed slides or cards as well as provide newsflashes, credits, titles, and other support information.

The TCG-225 produces one or two lines of 25 characters each. Character-generator electronics are activated by a keyboard similar to that of a typewriter. Besides containing alphameric symbols the keyboard controls every function of the system, including a unique hop-left and hop-right centering adjustment.

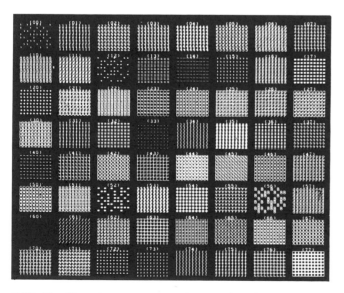

7-30. The Tarps language: characters produced on the S-D 4060 and textures used to shade an area.

VIDEO CHARACTER GENERATOR

The Aston **video character generator** and **multipage message store,** developed by the Denlen Electronics Corporation, is a system that can compose alphameric messages, store them in the computer's memory system, and superimpose them on programmed video signals—either internally within the unit or externally in a suitably keyed vision mixer.

VIDEO SYNTHESIZER

The **video synthesizer** is an analog-scan-conversion system that can convert the component part of an image into electric signals, modify the signals in compliance with the computer programmer's instructions, and display the continuously changing image on a monitor. The subject matter or input information for video-synthesizer systems may be either flat art, Kodaliths viewed by a TV vidicon camera, or the output from previously recorded videotape or processed motion-picture film. These images may be black-and-white or color, continuous-tone or high-contrast.

In addition to previously prepared artwork an infinite variety of abstract shapes, lines, and patterns can be generated within the synthesizer. The shape of these input images, whether previously prepared or newly generated, can be altered in part or in their entirety. Turning a dial or flipping a switch can change the height, width, shape, and position of the picture information visible on the display. The images may be rotated or zoomed from one position in the frame to another to create the feeling of depth.

The video synthesizer consists of units for controlling the graphic display along with modules for generating and controlling the animation. The display unit includes the height, width, depth, vertical, and horizontal positioning and centering controls and the intensity settings that regulate the brightness of the displayed image. The height control is capable of shrinking an image to a fine horizontal line, inverting it, and restoring it to its original position. Conversely, the width control can reduce an image to a thin vertical line or expand it until it fills the entire screen. The depth control causes the image to advance or recede, creating two- and three-dimensional effects. The vertical and horizontal positioning controls move the image up or down, to the right or left, and on or off the screen. The horizontal and vertical centering modes can move an image through a preset pattern and also make adjustments in either the horizontal or vertical axis.

The synthesizer's animation-control modules include a summing amplifier, diode module, ramp generator, audio interface, and wave-form generators. The summing amplifier is used to combine functions if programmed instructions call for a number of routines to be performed simultaneously. The diode module divides the generated wave forms and timing ramps that automatically control the preset speed and length of an animation sequence. The audio interface drives the animation by using audio signals. The wave-form generators, in addition to producing the graphics for display or for subsequent animation, control sync, frequency, wave form, and amplitude and frequency modulation.

Programmed movement may be rehearsed, and, if changes are considered necessary, the entire picture or any of its component parts can be retimed or reanimated. The subsequent real-time preview, or dry run, is precisely repeatable. Unusual illusory effects can also be produced by assigning specific synthesizer-controlling frequencies to certain sounds such as the notes of a musical instrument or the resonances of human speech. The sequences of motion, displayed on a high-resolution monitor, can be picked up by a TV plumbicon camera for immediate broadcast, photographed by a motion-picture camera, or fed to a videotape-recording system for use at a later date. See Scanimate.

VIDIFONT

Vidifont is a TV-display system that can produce word messages from a number of different fonts and sizes in real time. The Vidifont enables the user to produce more creative and informative video displays for television broadcasting, advertising, and film production. Developed by CBS Laboratories, the type-font characters have high resolution and maximum viewer readability. Language symbols and characters in Japanese, Hebrew, Greek, or Russian can also be used with this display system. Proportional character spacing is a key feature, and character-display color control is provided on a word-by-word basis.

7-31. The Chromaton 14 video synthesizer. Designed to produce animated video graphics in motion and color, the synthesizer accepts one or two monochrome TV signals. It can colorize black-and-white scenes in five discrete colors. Four invisible levels, similar to the animator's acetates and solid background, are simulated. Segments of the display obtained from the cameras and internal generators can be placed on any or all of the four levels. Each level and the background are assigned a color by the programmer. The level concept together with other features provides an electronic-animation capability: the unit can connect to any existing studio system using studio sync, or it can operate as a separate unit by connecting it to a color monitor through an accessory built-in digital color-sync generator. (Courtesy of B.J.A. Systems, Inc.)

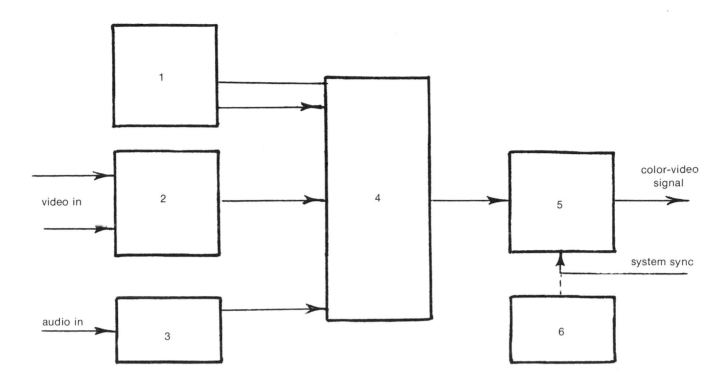

7-32. The Chromaton 14 video synthesizer: (1) pattern-memory generation and control, (2) input video, analog-to-digital converter, (3) audio processor, (4) digital processor, (5) digitally controlled color-video-signal generator, (6) sync generator (optional).

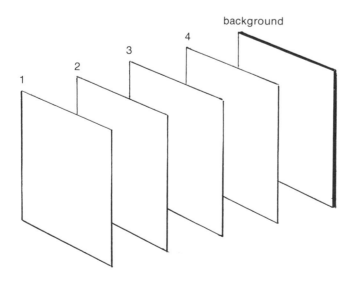

7-33. The level concept: segments of the display are placed on any or all of the four levels. Each level and the background are assigned a color by the programmer. Levels (1), (2), or (3) may be assigned that cel's color. Level (4) may be made to appear in front of level (1) but behind levels (2) and (3), producing weave effects.

COMPUTER OPTICALS

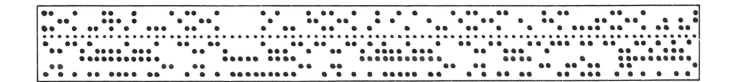

COMPUTER OPTICALS

Hunger-La Faim an animated film by Peter Foldes which was produced by a computer system developed in a collaborative effort by the National Research Council and the National Film Board of Canada, was awarded the Prix du Jury at Cannes in 1974 and fared equally well at festival competitions in other parts of the world. Nominated for an Academy Award in 1975, the animated short fell short. Unlike other nominees, the computer had not been programmed for an award-winning acceptance speech. *Norman Rockwell's World* was successful in its bid for the highly coveted Oscar. The director and the cameraman shared the credit with the electronic magic box. The next acceptance speech will probably be made by an optical cameraman.

Credit for the development of the electronic, computer-controlled printer must be shared by special-effects technicians and manufacturers of computer systems. The requirements of the first group were translated into specifications and prototypes by the second group and subsequently combined to produce a system that can achieve optical effects on film at speeds comparable to those of the tape medium. Some of the pioneers whose efforts helped turn the dream into a reality are Harvey Plastrik of Computer Opticals in New York, Charles Vaughn of Cinetron Computer Systems in Atlanta, and Edward Willette of Photographic Equipment Service in New Rochelle.

Because of the many additional controls required the production of a computerized optical printer presented a far greater challenge to the manufacturers than that of an automated animation stand. The problems associated with synchronization between camera and projector, nonexistent in computer-controlled animation-recording systems, had to be resolved before any of the related controls could be considered. Each problem was resolved—in turn.

Synchronization between camera and projector is absolute. Each unit is designed to function independently in either the forward or the reverse mode, in sync or out of sync. There is no mechanical linkage between camera and projector in the computerized printer. Absolute synchronization between both units is maintained by the computer itself, eliminating any chance of flicker. This type of electronic drive makes it possible to program any skip-frame or multiframe effect, with any mathematical combination, quickly and accurately. The printer also has a manual override for simpler effects. And because the computer can drive both the camera and the projector at much faster speeds than are obtained with the larger motors and clutches in conventional printers, electronic visual-

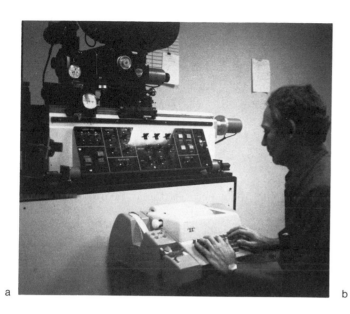

8-1. (a) The computer-controlled optical printer, teletype, and control-tape punch. (b) The teletype and control-tape punch. (c) The control panel of the computerized optical printer. (d) Loading film on the projection side of the printer.

display counters are incorporated instead of the mechanical type formerly used. As a result of these innovations the printer is able to operate on a complete program, locate scenes in the various projectors, bring the proper frame into exposure position in the camera, and carry out all of the preprogrammed instructions.

In practical terms the computer-controlled optical printer means that the producer or editor does not have to compromise on a desired effect because of the complexity of the operation involved. Before computers zoom or pan movements within any part of a 2"-×-2" transparency were limited not only by industry technology but, more importantly, by the cost factor. For the low-budget producer of slide-motion films optical moves constitute the major part of the optical bill. With the computer-controlled printer, however, the time factor involved in the production of zoom and pan effects is substantially reduced. A 10' zoom or pan move, for example, requires little more plotting and execution time than a 1' move. Movement within a transparency can be achieved as surely as an artist applies brush to canvas. Panning direction can be changed as often as desired with a smoothness of movement that was never possible with a manually controlled printer. Capable of performing 198,000 calculations per second with absolute precision, the computer-controlled printer can create incremental moves of any distance in any number of frames, and each move is precisely repeatable. Similar complex zoom and/or pan effects would be prohibitive in cost if not impossible to achieve with the mechanical counters that are used for measuring distance on conventional equipment.

The computer calculates image size by **percentage,** not by field size. A zoom started with the camera at its extreme position can be continued with the aerial projector to the final position with no apparent changes in speed at any point of the effect. This capability makes possible a much greater enlargement or reduction in the size of the image. And since there is no mechanical linkage between film plane and lens, focus is absolute at every incremental point. Similarly, computer light-valve control assures even density from one end of the effect to the other.

For the producer of a multimedia presentation (**videography**), the printer allows the combination of picture information and the creation of split-screen or montage effects on a single-strand 35mm negative. The multiple images are photographed with an anamorphic lens. There is no loss of picture area or image sharpness even if the effect is produced from original slides. When the processed film is subsequently projected through a standard 2:1 anamorphic lens, the composite image fills the screen with an aspect ratio of 3:1. An imaginative director can create zoom and pan moves within any of the slides used to form the split-screen effect, program transitions, or zoom one of the images up to full-screen size.

The fade-and-dissolve mechanism can produce transitional effects of any length from four frames to infinity. Shutter opening and closing is programmed to follow the shoulder-and-toe curve characteristic of the film in use. Unlike the conventional optical printer, which can produce only one shoulder-and-toe curve because it is designed to follow the contours of the guiding cam, the computerized effect is a smooth, flickerless scene-to-scene transition at any film length.

Programmed to respond to industry terminology, the computer can reverse a movement. Instructing the computer to flip-dissolve, for example, reverses the direction to produce a dissolve from out to in rather than from in to out—the exact opposite of the previous command. The same command can be used to reverse any effect previously typed into the computer.

Other transitional effects such as wipes or scene pushes can also be accomplished in any length quickly and precisely with no riding between scenes. Mattes are no longer necessary to achieve an absolute butt (a nonfluctuating line) between incoming and outgoing scenes. Starting at any specific point, the computer repeats each of the moves in the effect with absolute precision as many times as is required. This capability assures matched moves on insert shots and background pans.

A motorized lens compound driven by the computer maintains camera and aerial-projector focus at all points of a zoom movement from reduction through blowup. In order to utilize the most efficient f-stop of any selected lens, including anamorphic lenses for wide-screen formats, a computer-controlled light valve, which maintains a constant level of light regardless of camera or

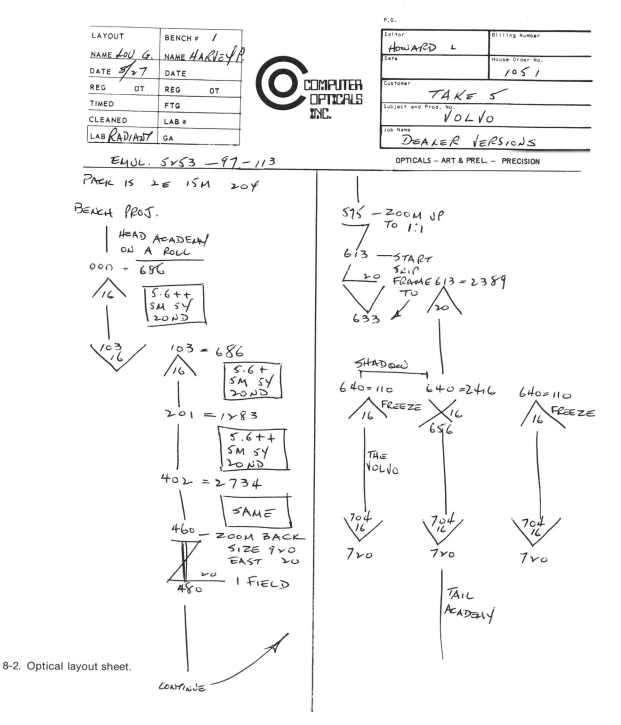

8-2. Optical layout sheet.

135

```
ZEROQU,103,32,29,29
QU,201,32,29,29
QU,402,32,29,29
QU,613,31,28,28
CU,0,0,0
TIP
SYP
AD,0,686,0
PUN
FA,0,16,IN
SH,103
PU
DI,103,16,0U
SH,119
PU
AD,103,686,0
PU
FL,DI
SH,201
PU
AD,201,1283,0
PU
SH,402
PU
AD,402,2734,0
PU
SI,920,460,480,3,BO
EA,20,460,480,3,BO
SH,575
PU
SI,1000,575,615,3,IN
WE,20,575,615,3,IN
DI,613,20,0U
SH,613
PU
SK,1,1,633
PU
AD,613,2389,0
PU
FL,DI
SH,640
PU
DI,640,16,0U
SH,656
PU
AD,640,2416,110
PU
UNTA
FL,DI
SH,704
PU
FA,704,16,0U
SH,720
AD,640,3200,0
PU
CU,640,0,0
AD,640,110,0
PU
UNTP
FA,640,16,IN
SH,704
PU
FA,704,16,0U
SH,720
EN
```

aerial-image position, is incorporated in the design of the printer. A color-additive light source makes color corrections on a scene-to-scene basis, eliminating the need for corrective filters—a big plus for the computer-controlled printer.

Like the more conventional models, the computer-controlled printer accepts both 16mm and 35mm transport mechanisms. A provision for interchanging shuttles and sprocket assemblies makes it possible to handle any combination of raw stock and previously processed film—16mm camera stock with 16mm elements on the projector side of the printer, 16mm camera stock with 35mm elements in the projector head, 35mm raw stock with 16mm projector elements, and 35mm raw stock with 35mm elements in the projector. An added advantage not shared by conventional printers is the ability to make major changes during the actual filming by simply updating the program.

The basic command language for the computer-controlled optical printer is similar in many respects to that for the computerized animation stand (see Cinetron). The principal differences between the two computer languages center around the controls that are used mainly to achieve optical effects, some of which are described below.

8-3. Programmed instructions. (Courtesy of Computer Opticals, Inc., New York.)

8-4. Punched control tapes.

COLORHEAD CONTROL

The command co, (red value), (green value), (blue value) sets the color-additive head so that specific color corrections can be programmed in any combination of primary-color values. This capabilty completely eliminates the use of corrective filters. The instruction for setting the color-additive head to 25 red, 30 green, and 45 blue, for example, is written as color, 25, 30, 45.

QUICKCODE

The command QU, (red value), (green value), (blue value), (target frame) is used if a color change is required in only one scene or in a stated specific area of the film being duplicated. The instruction has no effect on the colors in other areas of the running footage in the printer's projection head. All other conditions of interlock and projector counts remain unaffected. The target frame represents the final frame count for the indicated color change.

The command serves as a terminator for the input, so the printer begins to run as soon as the new instruction is typed. If the final frame for the correcton was 3445 and the color-additive-head settings were similar to those used in the colorhead-control example, the instructions would be typed as qu, 25, 30, 45, 3445. This typed command would have exactly the same effect as writing the group of instructions co, 25, 30, 45; 3445; run.

SYNC

This command causes the aerial or projector heads to operate synchronously with the camera. Sync aerial (sya) and sync projector (syp) are the instructions for the specific modes.

UNSYNC

The unsync (un) identifier causes the specified head to proceed to the 180°, or freeze, position. It also takes the programmed head out of interlock to prevent accidental exposure or film advance. Interlock (in) is established between the camera and the projectors by using the identifier tie (aerial or projector). The cameraman types the identifier and names the head, as in the sync and unsync routines.

ADVANCE

The instruction ad, (camera target), (projector target), (aerial target) causes the printer to advance at rewind speed (720 frames per minute) to the programmed frame counts on the indicated heads. The command to run to 200 camera, 14400 main, and 1332 aerial would be typed as ad, 200, 14400, 1332; run. The computer awaits further instructions after the indicated counts are reached.

CURRENT COUNT SETTINGS

The programmer or cameraman can set the computer frame counts for the various heads by using the current (cu) instruction. The frame counts for 200 camera, 300 main, and 500 aerial would be typed as cu, 200, 300, 500. This command causes the computer to set the numbers as the current values for the specified heads. The heads themselves do not move until the next programmed routine orders a specific operation.

SKIPFRAME

The typed instruction sk, (frames interlocked), (frames skipped), (final camera count) orders the camera to skip every other frame until the required frame count is reached. It is typed as sk, 1, 1, 2223. The instructions for skipping every third frame is typed as sk, 2, 1, 2223.

MULTIPRINT

The typed instruction mu, (number of hits per frame), (final count) is used to program multiple printing or duplication of the same frame of film. The instruction mu, 2, 2223 would initiate a double-printing routine to frame 2223.

The use of a special routine does not affect other instructions. A simultaneously programmed effect such as a 16-frame fade, dissolve, zoom, or pan would be carried out over the specified skip-frame or multiprint footage.

VIDEOTAPE

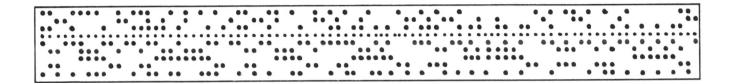

Magnetic sound recording is generally regarded as a relatively recent development, but in fact, its origin can be traced back to the 1880s and a Dane named Valdeman Poulsen. The first of the crude devices used to record sound on lengths of steel wire produced equally crude sounds during playback. The distorted output began to clear up in 1928 with the development of a magnetic material, iron oxide, that could be applied as a coating to a paper-tape base. A device capable of carrying both picture and sound information on the same magnetic surface was the next logical step: the first commercial **videotape recorder** (VTR) was announced by Ampex in 1956.

The sophisticated equipment currently available from manufacturers such as Ampex, CVI, EECO, IBM, RCA, CBS, 3M, and Sony incorporates features designed to eliminate tedious, repetitive, time-consuming manual operations; increase efficiency; and reduce overall production costs. Compared with the relatively simple formats in the motion-picture industry—35mm film, for example, is exposed as it passes through 35mm shuttles in 35mm cameras, and 16mm film requires 16mm equipment for exposure and processing—videotape formats vary considerably. It is produced in a number of widths (2″, 1″, 3/4″, 1/2″, and 1/4″) for either **helical** or **quadruplex** recording, and sound and picture information can be recorded at different speeds (15″ per second (ips), 7 1/2 ips, and 3 3/4 ips). At this time quadruplex recording on 2″ tape is the broadcast standard for the industry, although it is difficult to discuss standardization in an industry in which the very material used for recording is available to the consumer in so many different forms. The same tape, in each of the various widths and with the same iron-oxide magnetic particles impregnated in the surface, is available on open **reels** for reel-to-reel recording; in **cassettes,** on which the tape travels past the erase, record, and playback heads from the feed reel to the take-up reel; and in **cartridges,** which have single reels. Fortunately, electronic signals recorded in one format can be reproduced for playback in another format simply by rerecording the original picture and sound information. Since only a negligible loss of quality is involved in the dubbed copies, the problems associated with equipment standardization and interchangeability are not insurmountable. One of the most popular compromises at the present time is the 3/4″ U-Matic color videocassette system.

In sound-recording equipment record, erase, and playback heads in the transport mechanism generate, store, and reproduce the audio signals recorded on the magnetic tape. The general range of recording frequencies, approximately 20 to 20,000 hertz, parallels the range of human hearing—

a hertz (hz) unit is equal to one cycle per second. Although most types of commercially available equipment can record sound at a number of tape speeds (15″, 7 1/2″, 3 3/4″, and less per second), the best fidelity at higher frequencies is obtained with the fastest recording speeds. By comparison, the video signal requires frequencies up to 5 million hz, which in turn require proportionately faster recording and playback tape speeds—a minimum of 500 ips, for example. Since these velocities are impractical, alternative methods in which the tape passes over rapidly rotating record heads compensate for the greatly reduced tape speeds (see helical-scan and quadruplex recording in the computer glossary).

The electronically encoded images produced on videotape for subsequent broadcast in the United States cannot be reproduced over European, Russian, African, or Australasian networks. The American 525-scanning-line format and 60-hz electric current are incompatible with the 625-line, 50-hz standard used on other continents. Both systems, however, accept any of the three types of videotape recordings—reel-to-reel (open reel); cassettes, with built-in feed and take-up reels; and cartridges, housing a single reel of tape that carries both picture and sound information within their own standards. The old adage, "If one is good then two must indeed be better," is confusing in the case of videotape standardization and related film-production techniques. The producer, for example, can complete all of the production and postproduction processes on film and then have the composite duplicated and distributed on videotape. Similarly, he may choose film for the actual production and have the picture information transferred to videotape for postproduction editing. The alternatives are many.

VIDEOTAPE-TO-FILM CONVERSION CHART

tape frames	film frames
1	1
2	2
3	2
4	3
5	4
6	5
7	6
8	6
9	7
10	8
11	9
12	10
13	10
14	11
15	12
16	13
17	14
18	14
19	15
20	16
21	17
22	18
23	18
24	19
25	20
26	21
27	22
28	22
29	23
30 = 1 second	24 = 1 second

VIDEOTAPE RECORDING

Videotape recording combines the best features of two media, film and television. The versatility of the film medium coupled with the presence and immediacy of live television provide the viewer with the most effective means of communication ever devised—and it is electronic.

While the staging techniques for motion-picture and videotape productions are basically similar, the hardware is quite different. The film producer, for example, must adjust his thinking to the fact that the television camera focused on the scene contains no film. The second shocker is the fact that the magnetic tape threaded in the videotape recorder has no frame lines separating one invisible image from the next. And although a combination of optical, mechanical, and chemical processes is needed for motion-picture production, television and videotape recording and transmission techniques require optical and electronic processes only. Since the latent image does not have to be developed in the laboratory, picture information is available for viewing on studio monitors as soon as the recording process is completed. This capability not only speeds up production but also completely eliminates the need for insurance takes and the ulcer-producing waiting period that precedes the viewing of motion-picture dailies, or rushes. The possibility of immediate playback and preview of the recorded picture information places the producer, director, performer, and technician in a more relaxed atmosphere. Corrections and revisions are made immediately, usually on the same length of tape; flaws in performance that escape detection during taping are certain to be spotted before the set is struck; script revisions, lighting changes, and new camera angles may be planned immediately. In most cases rerecording takes less time than the discussion that precedes it.

The hardware for the basic videotape-recording system consists of one or more television cameras, videotape recorders, and television monitors. The monitors are connected directly to the television camera or videotape recorder and can display only the electrical signals received from the particular unit. A television receiver, by compari-

son, can also receive and reproduce standard broadcast signals. Portable systems such as Porta-pak are relatively light (approximately 20 to 30 pounds), battery-powered, self-contained units. These systems include a small television camera with an electronic viewfinder and built-in microphone, a control unit, and a videotape recorder—don't forget the batteries.

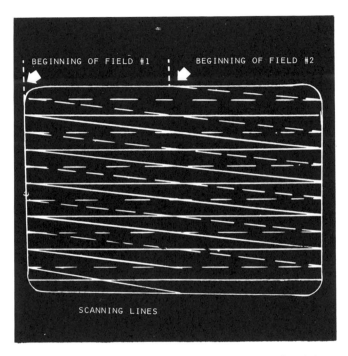

9-1. Interlaced scanning. The electron beam traces the picture on each of 525 scanning lines to form two interlaced fields, or one complete video frame.

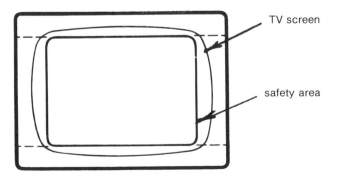

9-2. Cutoff area.

140

The role of the motion-picture director on a set or on location and that of the videotape director during a recording session differs considerably. The motion-picture director plans closeups, cover shots, inserts, and visual effects in addition to supervising the actual filming of the scene. The cover shots are designed to provide the film editor with a greater degree of flexibility during the post-production stage. On the creative end of the videotape-recording process, however, the director and the editor share responsibility and concentrate on the output image displayed on studio monitors. The relatively small filming crew includes a camera operator, a recordist, who operates the videotape recorder, and a video colorist, who is responsible for light-level adjustments and primary-color balance. If two cameras are used during the same session, the output from each is visible simultaneously on monitors in the studio control console. Editing decisions are made after the recording process is completed.

Recording generally centers around a studio-type, broadcast-quality color camera such as the RCA TK44, RCA TK45A, or Norelco PC-70. Smaller models, including hand-held cameras such as the RCA TKP45 or the PCP-90, are also used if necessary. Angenieux, Canon, and Schneider zoom lenses are used most frequently for studio and location filming assignments. Fixed focal-length adaptors for electronic TV cameras make it possible to use standard 35mm cine, 35mm single-lens-reflex, and special-effects lenses such as the fisheye with a minimum of light loss or change in the angular field of view. Lighting problems are virtually nonexistent. Videotape requires considerably less light than that needed on the motion-picture set: the usual **key-to-fill ratio** is approximately 2:1. Film has a reproducible **gray scale** of 1 to 100 (whitest white to blackest black); the videotape gray scale, by comparison, is only about 30 to 1.

In videotape recording what you see is what you get. Camera moves, lighting effects, color balance, and the performance itself are monitored by the director while the scene is being recorded on videotape. Seated at the control console, the director evaluates the output, and any revisions and/or corrections are made immediately after the instant replay is reviewed. The waiting period that precedes the screening of dailies and the subsequent striking of the set in motion-picture production is nonexistent in videotape-recording processes.

VIDEODISK

A **videodisk** looks like an overweight phonograph record. The transparent plastic coating supports both a picture and one or more sound tracks. The disk is capable of reproducing visual information via existing components and hardware such as a standard television receiver. There are two basic types of videodisk-recording systems—the Philips-MCA and the RCA Selecta-Vision. Other disk-recording systems offer only minor variations from the basic techniques described below.

Unlike a videotape, with its iron-oxide layer on which picture and sound information is translated into magnetic impulses, the Philips-MCA videodisk images are recorded, stored, and replayed as a se-

9-3. The videodisk player for the Philips and MCA videodisk systems is easier to operate than a conventional phonograph and simple and safe enough for a child to handle. Once the videodisk is in place, a touch of the play button starts the programming. The disk itself is more durable than a long-playing record, and since there is no physical contact between the player's laser readout and the videodisk, playback quality remains unchanged even after years of use.

ries of light patterns. During the recording process, which relies on light refraction, or bending, sound and picture are converted into signal waves, or pulses. Each of the pulses etches a minute depression into the surface of the disk's photographically sensitive coating by means of a laser system that focuses a pinpoint of light proportional to the strength and magnitude of the input signal. The depressions form a spiral pattern analogous to the grooves in a conventional phonograph record. During the playback mode the recorded disk, with its protective, transparent-plastic coating, is placed facedown on a turntable. The depressions etched during the recording process are scanned by a focused laser beam, reflected to a photocell, and reconverted into picture and sound information.

The playback unit is simply hooked up to the antenna terminals of a monitor or home receiver, and the system is go. In the Philips-MCA videodisk system the combined video/audio wave forms vary in length from 0.8 to 2.5 microns (a micron equals 1/25,400"). The length of the wave form is constant at approximately 0.7 microns. The depth of the depression formed during the recording process is 0.1 micron. Three million depressions, or bits, are recorded per second and packed into grooves about 1/13,000" in diameter. The standard phonograph record, by comparison, has between 150 to 200 grooves to the inch. The disk spins at the rate of 1,500 rpm, and each revolution carries a single frame of picture. The disk provides one-half hour of program material.

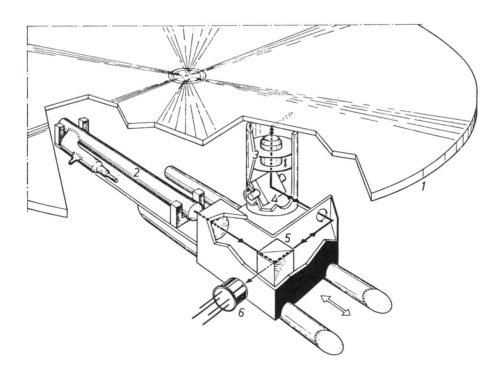

9-4. The VLP (video-long-play) television-disk system: (1) The VLP record, (2) he-ne laser light source, (3) lens assembly, (4) galvanometer mirror, (5) beam-splitting prism, (6) photodiode (detector). (Courtesy of Philips Research Labs, Eindhoven, The Netherlands.)

In the RCA system the picture and sound information is fed into an electron gun and converted into a stream of electrons by means of a powerful microscope. The electrons are projected onto the master recording disk and interact with the sensitized surfaces that cover the millions of mechanically cut indentations in the copperplate. The modulated electron stream that represents the input video signal is recorded on the disk's sensitive layer. The minute electrical differences in signal strength are pressed into the indentations and duplicated on vinyl disks covered with a metal surface designed to increase the conductivity of the indentations. The pressing is finally sealed with a lubricant that protects the disk and increases the life of the stylus that tracks along the grooves formed by the indentations. The disk rotates at a speed of 450 rpm, has 5,555 grooves per inch, and provides four video frames per rotation. Both sides of the disk can be utilized to provide a full hour of picture and sound information.

Magnetic disks that use rotational pieces of tape for recording picture and sound information are gradually being phased off the market, leaving the field to the mechanical and optical systems. Other systems in various stages of development include a glass disk coated with a thin metal film that is etched with laser beams, a disk that utilizes photographic principles, and a transparent disk with laser-beam readout. Videodisk recordings enjoy a distinct advantage over other recording systems because of their random-access capabilities.

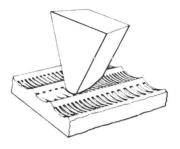

9-6. The RCA Selecta-Vision videodisk system. The picture and sound information is contained within the millions of incised indentations at the bottom of each tracking groove—there are 5,555 grooves to the inch. The stylus travels over the disk surface as it rotates at 450 rpm, providing four video frames per rotation.

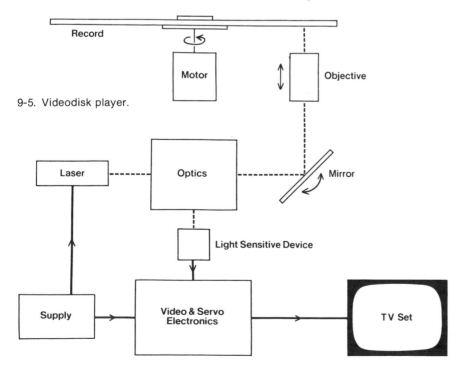

9-5. Videodisk player.

ELECTRONIC EDITING

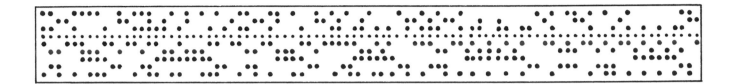

Videotape-editing techniques until recently were analogous to conventional procedures—assembling individual scenes; synchronizing them with the dialogue, music, and effects tracks; and guiding the many elements through the mix and optical processes needed to produce a smooth, continuous flow of picture and sound information. But the inherent speed, versatility, instant-replay capability, and flexibility of the newer computer-controlled electronic-editing systems have changed the state of the art. Today's electronic systems eliminate tedious, repetitive manual operations, increase director-editor efficiency, and drastically reduce production costs.

By any other name—the following electronic-editing systems are all designed to transfer the contents of a roll of recorded videotape to some form of storage facility from which segments can be accessed quickly and to assemble the retrieval information into a smooth flow of visual continuity in sync with the accompanying sound.

CDL, an acronym for Central Dynamics Ltd. of Canada, is a completely computerized system. Operating under software control, it performs the basic editing operations of synchronization, sequencing, storing, and switching.

CMX is a computerized-editing system developed by CBS and the Memorex Corporation.

Edicomp is an electronic-editing system in which a programmer accumulates the editing data that is subsequently used to transfer the original sound and videotape picture to an edited second-generation copy.

Editec is an Ampex system in which the edited work print is used as a guide for the rerecording process and for the production of the composite videotape.

EECO is an acronym for Electronic Engineering Company of California, a manufacturer of broadcast products and equipment for electronic-editing systems.

RA 4000 is a computerized system used in videotape editing.

RCA T.C.E. is the RCA time-code editing system.

RCA T.E.P. is an acronym for tape-editing programmer, an electronic-editing accessory used for scene-by-scene reassembly of original production tapes.

Using simple, logical language and response, the computer-controlled electronic-editing system's mathematical routines, data-entry/error control, and transport mechanisms extend the manipulative skills of the director and the editor. The electronic systems's design encourages creativity and spontaneity. All decisions are critically reviewed before they are executed. Designed to reject invalid instructions, the system brings questionable decisions to the editor's attention for further review. The interrelationship between the programmer and the control console requires a minimum of effort. Communication is accomplished via the control-panel keyboard and the CRT display at the control console. Electronic-editing systems incorporate and utilize the SMPTE **time-and-control code**, which is impressed on each frame of the videotape. It permits rapid and accurate information retrieval, editing, assembly of programs, and verification of tape content.

The system brings the advantage of real automation to the videotape-editing room in a new, flexible format. The processing power of the digital computer is teamed with an optimized, interactive **command/display console** to allow the director and editor artistic and creative freedom, since all checks, mathematical calculations, and operations are performed automatically and rapidly. Data verification, operational monitoring, and system-configuration routines guarantee reliability during every phase of the editing process.

Electronic-editing systems consist of modular building blocks with discrete but interrelated functions. The command/display console (CDL's PEC-102, for example) serves as the communication center and displays all information, sequential operations, and instructions to the programmer graphically on the CRT. Redundant information that might constrain or confuse the director or editor is not included in the display. The software—the system's brain—is the computer. The computer compiles, memorizes, stores, and retrieves all information and data and executes the programmer's commands immediately and precisely. The computer's magnetic-disk memory has a capacity of 64,000 computer words and can store up to 600 program-edit scenes. The **edit-file disk** can be loaded and controlled from the console or loaded with program-scene sequences through the teletype by means of punched-paper control tapes. The teletype is the direct link between the programmer and the command/display console. A typical automated-editing session consists of several basic modes of operation, initiated by the software. Any incorrect operating instruction during any part of the operation results in a no-go condition, which is highlighted by a flashing message.

The **initialize** mode in the electronic-editing system configures the mechanical system to be used: it identifies and defines the equipment assigned to the operation, the mechanical operating parameters, and the form of address for the videotape recorders, Slo-Mos, multitrack audio-tape decks, helicals, film projectors, and studio cameras. The **tape-load** mode identifies the tapes to be used and specifies the exact length of the final edited program. The system's **time-code reader** identifies the information (hours, minutes, seconds, and frames) encoded on each frame of the videotape by the SMPTE/EBU time-code generator. The **audio/video machine interface** ties all the information, command, and decision units into a compatible, immediately accessible system. In the **operate** and **machine-control** modes the exact time-code position of equipment assigned to play, fast-forward, or rewind is displayed on the control console. The system accepts the programmer's commands, interprets and calculates durations and edit-in and edit-out addresses, and updates scene sequences. All video and audio effects are programmed along with the length of the desired transition. The video effects available at the editor's fingertips include wipes, dissolves, matte keys, and split-screens. In the operate mode the system recues and synchronizes the machines and aborts an edit if necessary. Audio and video sources on the same tape can be offset and recorded in one pass. This mode also automatically monitors and transfers relevant scene-edit information to the edit-file disk. The **video/audio switcher**, an auxiliary unit equipped with a fully manual control panel to accommodate off-line production work, adds the video schmaltz and the audio sweetening. Another accessory, the **autoframe**, provides the all-important color-phase detection and correction capability. The **time-code inserter** displays the time code with the video to provide visual identification of individual frames for off-line editing. The console's character-display generator has a tremendously fast data-transfer rate—60,000 characters per second. This capability is required in order to display the time-code position of each of the interconnected units instantaneously during all operational modes. If necessary, machine time-code-address positions can be distributed to remotely located equipment for display on interconnected monitors. The search routine manipulates 600 scenes on the edit-file disk and displays this data base on the console CRT along with the scene number, record-machine position, edit-in and edit-out address for each segment, tape number, and desired transition. It can delete or insert a scene and automatically calculate the new time-code addresses. If required, the disk can be loaded with a new sequence from a punched control tape at any time during the assembly process.

HARDWARE AND SOFTWARE

The audio/video switcher adds the video schmaltz and the audio sweetening to the postproduction process. The switching unit is responsible for programming and initiating effects and transitions; it also generates soft-edge wipes and chroma-key effects. The module is equipped with a fully manual control panel to accommodate off-line production work.

The autoframe detects and corrects color phases during editing. It is required on the record machine only and interconnects to the machine reference pulses, video output, and servo system. The best edit-timing stability for the record-machine editor is achieved by utilizing the unit in conjunction with a processing amplifier, which is used at the input to insert sync at a fixed timing into the video.

The software—the electronic-editing system's brain—is the computer (the CDL PEC-102 is described here). It compiles, memorizes, stores, and retrieves all programmed information and data. The mathematics for all of the individual software programs—initialize, tape load, operate, machine control, and search—are calculated and controlled by the computer. The main memory is augmented by the edit file, which can store up to 600 program-edit scenes. This fixed magnetic disk is tied directly to the rest of the system and to the teletype. The 50-×-16-character CRT display generator allows updating at maximum rates in order to present the current time-code addresses in the various units that are tied into the system. Three time-code-recovery units, designated as record, playback 1, and playback 2, are assigned to the machines. The interface to the audio/video switcher allows the program to select cross points, transitions, and effects; set up transition rates; and start transitions in both audio and video sections. Data input and output for the system is handled by the teletype, which provides both hard-copy and paper-tape operations.

The cue module, added to the thumbwheel-switch module, provides rapid, automatic search for the initial edit point plus accurate preroll parking at a predetermined location ahead of the initial point.

The dual-cue controller starts two tape recorders (either audio or video) simultaneously or individually in the proper direction and stops the machines at the selected cue points. Cue points are selected by either front-panel or remote switches. A microcomputer in the control erases the cue points and the SMPTE edit code, decoded from the recorder tapes, to calculate the directions and distances to the cue points. Digital-to-analog converters and control circuitry provide analog voltages to drive the recorder reel motors.

The edit-code generator (the EECO BE520 unit is described here) develops SMPTE/EBU edit codes that are used for electronic indexing and editing of video and audio tapes. The output code can be used as a real-time event or as an arbitrary elapse-time index for either NTSC or PAL television standards. The edit code is recorded on magnetic tape and precisely identifies, in an 80-bit binary message for each frame, the hour, minute, second, and frame number. The message also provides digits for extra unassigned information (referred to as user bits). The time code can be synchronized to either line-power frequency, composite video, or composite sync. It can be recorded on magnetic tape on either audio or video machines to provide a controlling signal for cueing an audio or video recorder, synchronizing a recorder to another recorder, or editing audio and/or video from playback to record machine. The nixie display shows the generated time of day, or elapsed time, from 0 to 24 hours. The displayed time is available at an output connector as parallel binary-coded decimal signals. These signals can be used to drive auxiliary time displays, event-control devices, or video character generators

such as EECO's BE400, which develops identifying television signals corresponding with the indexed hour, minute, second, and frame information.

The edit-code reader demodulates and decodes the SMPTE/EBU edit code and visually displays the time and frame count on an eight-digit readout. The input code can be from any source, such as audio and video tapes or the edit-code generator. The display is presented as a reference during tape

10-1. The EECO BE520 edit-code generator. (Courtesy of Electronic Engineering Company of California.)

10-3. The EECO BE 421 edit-code reader. (Courtesy of Electronic Engineering Company of California.)

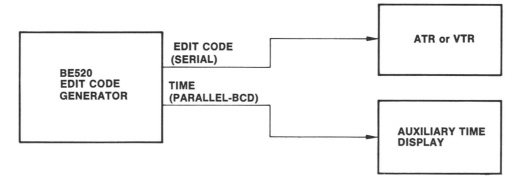

10-2. The EECO BE520 edit-code generator develops the SMPTE/EBU edit code, a serially formatted code consisting of 80 bits of binary-weight data. The generator provides precise synchronization of the code with either power-line-frequency or video-sync inputs. The unit is compatible with NTSC and PAL TV standards. The generated time of day is displayed on the front panel of the unit. In addition to the hours, minutes, and

seconds frame numbers are also generated. The frame rate is the repetition rate of the edit code. The edit time code, recorded on magnetic tape (either audio or video), provides a controlling signal for cueing an audio or video recorder, synchronizing a recorder to another recorder, and editing audio and/or video from playback to record machine. (Courtesy of Electronic Engineering Company of California.)

147

play, rewind, fast forward, or single-frame hold. Selected scenes from either quadruplex or helical tapes can be reviewed and a program sequence developed to show the start and stop times. The program sequence provides an accurate index for manual, automatic, or computer-automated editing. Some readers contain look-ahead electronics that serve as a supporting accessory to the series edit programmer. Repeated reviews for electronic editing are expedited if fast and accurate cueing operations can be made, and the look-ahead logic, connected to the edit-programming modules, provides this capability. Parallel edit-code outputs from the reader can be used as inputs to the wide-range synchronizer, video-character generator, and other event-control devices. The unit automatically reads and displays the code information, whether it is received in the forward or reverse direction or in normal playback, reduced, or fast tape speeds. At any desired instant the programmer may stop the display from updating to hold the time and frame count in the display as an aid in scene identification. The frame count is displayed only during the hold condition: upon release the time display again updates to the time of the incoming edit code. Once it is logged, any scene can be located quickly for inclusion in a live broadcast.

The edit-code restorer is used to improve the quality of the SMPTE/EBU edit code when the code is dubbed from one tape to another. In dubbing tapes with time-code indexing the restorer should be electrically placed between the normal-playback amplifier and the output connector. This unit is attached to both audio and video tape recorders and operates at playback speed only.

The edit-file system is the storage facility for electronic-editing systems. The magnetic fixed disk, with its capacity of 64,000 computer words, can store up to 600 program-edit scenes. The disk is tied directly to the teletype and to the other units in the configuration. The edit file can be loaded and controlled from the console or loaded with program-scene sequences through the teletype with punched-paper control tape.

The editor-control module is used in conjunction with the thumbwheel-switch module and the cue module that controls the record VTR to provide pushbutton control of the preview and edit functions. The editor-control module regulates other modules so that editing is completed exactly as previewed—accurate to a single frame.

The building-block concept of the videotape indexing and editing modules contained in the edit programmer provides for the integration of peripheral equipment such as the edit-code reader, display units, switching units, edit-code generator, and video editing equipment. Edit points are located automatically, and all arithmetic calculations are handled by the programmer. Playback tapes are automatically slaved to the record tapes for start, synchronization, and stop so that outgoing material is delivered and received as scheduled with accuracy to a single frame.

The gen-lock board permits an operator using a special-effects generator to switch or achieve effects between cameras and a composite source such as a videotape recorder for a higher degree of program continuity.

The special-effects generator enables the programmer/editor to add wipes, split screens, and special effects such as mattes. It can be used for color or black-and-white production, and switching is done in the vertical interval.

The synchronizer module, added to the thumbwheel-switch and cue modules, provides automatic control of playback VTRs. The synchronizer module slaves the playback VTR to the record VTR and automatically brings the playback tape into sync with the master tape before the initial edit points are reached.

The thumbwheel-switch module is used to program accurate edits on both the playback VTR and the record VTR. The edit points for a playback tape or for the master tape are entered by setting the start frame number on the module's top thumbwheel switches, and the stop frame number on the bottom switches. By coupling the thumbwheel-switch, cue, and synchronizer modules playback tapes are automatically slaved to the record tapes for start, synchronization, and stop. Outgoing material is delivered and received as scheduled with accuracy to a single frame. The module allows fast and accurate preroll parking for preview and edit.

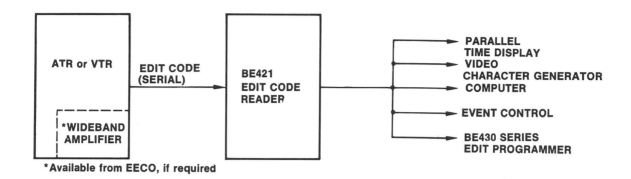

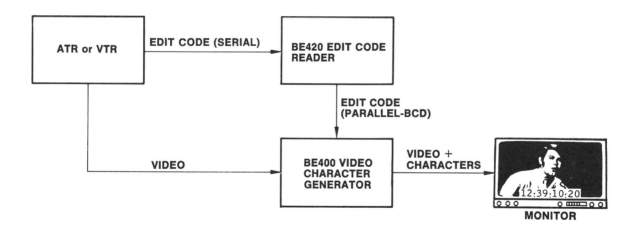

10-4. The EECO BE421 edit-code reader features visual display of decoded time received in the incoming SMPTE/EBU edit code. The unit is compatible with audio and video tape systems that employ time-code indexing. The edit-code reader is designed as a supporting unit to the edit programmer. Repeated previews for electronic editing are expedited by fast and accurate cueing operations. The look-ahead logic, connected to the edit-programming modules, provides this capability. (Courtesy of Electronic Engineering Company of California.)

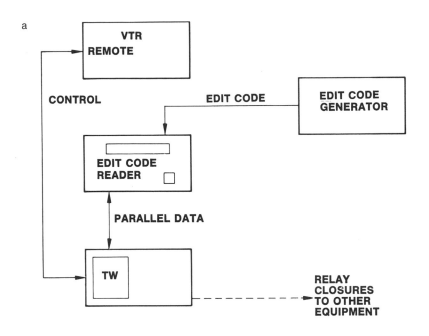

a

VTR
REMOTE

CONTROL

EDIT CODE

EDIT CODE
GENERATOR

EDIT CODE
READER

PARALLEL DATA

TW

RELAY
CLOSURES
TO OTHER
EQUIPMENT

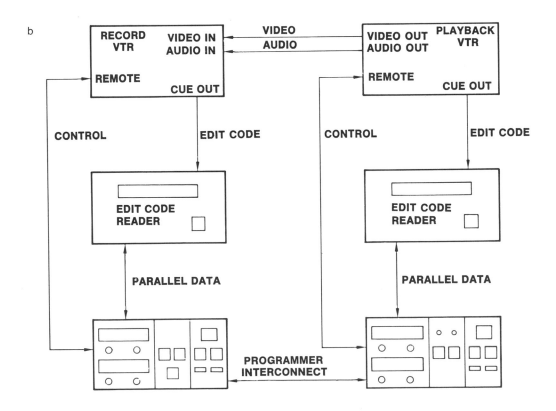

b

RECORD
VTR

VIDEO IN
AUDIO IN

REMOTE

CUE OUT

VIDEO

AUDIO

VIDEO OUT
AUDIO OUT

REMOTE

CUE OUT

PLAYBACK
VTR

CONTROL

EDIT CODE

CONTROL

EDIT CODE

EDIT CODE
READER

EDIT CODE
READER

PARALLEL DATA

PARALLEL DATA

PROGRAMMER
INTERCONNECT

The time-code inserter is a standard option for transferring the time-code read from a cue track or a time-code generator into the video of another tape that is to be used for off-line editing.

The purpose of the video character generator is to develop video characters that identify a specific television signal by the hour, minute, second, and frame. The video characters are generated by extracting the time and frame information from the SMPTE/EBU edit code on the videotape and by direct conversion of a parallel edit-code input. The video character generator can be used to superimpose the edit code on the video for display on picture monitors without affecting the tape being recorded. This feature allows the viewer to correlate specific scenes and information directly from the picture monitor for off-line editing.

The wide-range synchronizer is designed to slave audio tape recorders to quad-track and slant-track video recorders or to other audio tape recorders. Synchronization between sprocketed and non-sprocketed equipment is essential in the editing process. The synchronizer compares identical SMPTE edit codes from two tape sources and develops a phase-error signal, which is used to regulate the drive speed of the slave unit.

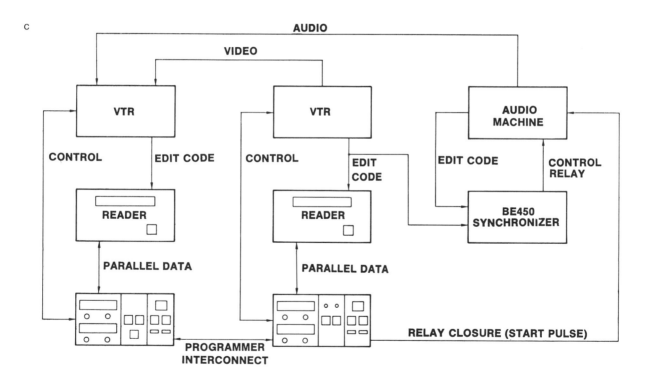

←→

10-5. The EECO series of videotape indexing and editing modules is specifically designed for small stations or budget-minded production facilities. The building-block concept allows the user to select only modules that he needs. The combination of modules in a single chassis controls the recorder—one chassis for each recorder in the edit system. Circuits are provided in the modules for integrating peripheral equipment such as edit-code display units, switching units, an edit-code generator, and video edit equipment. Regardless of the system configuration each edit programmer must be accompanied by an edit-code reader. Three interface configurations are shown here: (a) automatic start-stop operation, (b) two-machine operation, (c) dual-system operation. (Courtesy of Electronic Engineering Company of California.)

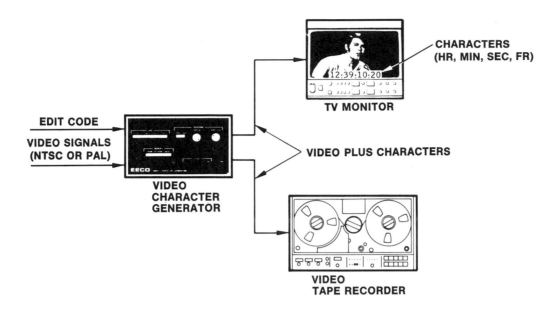

CHARACTERS
(HR, MIN, SEC, FR)

TV MONITOR

EDIT CODE

VIDEO SIGNALS
(NTSC OR PAL)

VIDEO
CHARACTER
GENERATOR

VIDEO PLUS CHARACTERS

VIDEO
TAPE RECORDER

10-6. The purpose of the EECO BE400 generator is to develop video characters that identify a specific television signal in accordance with the hour, minute, second, and frame. The video characters are generated by extracting the time and frame information from the SMPTE/EBU edit code on the videotape or by direct conversion of a parallel edit-code input. As an aid to editing the video-character generator can be used to superimpose the edit code onto the video for display on picture monitors without affecting the recording or broadcast. This feature allows the viewer to correlate specific scenes and information directly from the picture monitor during periods of off-line editing. The generator inputs are edit code and video. The edit code contains the time (hour, minute, second) and the video frame number. Generator logic converts the binary edit code and frame number into video characters (word and frames). The characters are then inserted into the video for presentation on a monitor or for input to a videotape recorder. (Courtesy of Electronic Engineering Company of California.)

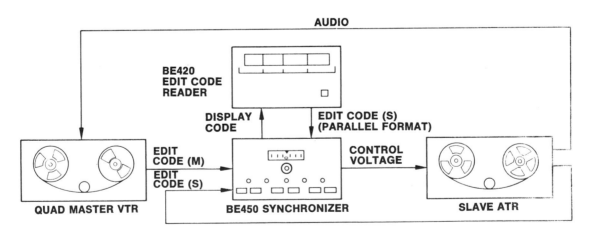

AUDIO

BE420
EDIT CODE
READER

DISPLAY
CODE

EDIT CODE (S)
(PARALLEL FORMAT)

EDIT
CODE (M)

EDIT
CODE (S)

CONTROL
VOLTAGE

QUAD MASTER VTR

BE450 SYNCHRONIZER

SLAVE ATR

10-7. The wide-range synchronizer is designed to slave audio tape recorders to quad-track and slant-track video recorders or to other ATRs. Synchronization between sprocketed and non-sprocketed equipment is essential in the editing process. The synchronizer compares identical SMPTE/EBU edit codes from two tape sources and develops a phase-error signal, which is used to regulate the drive speed of the slave unit. (Courtesy of Electronic Engineering Company of California.)

CDL SYSTEM

The modular building-block contruction of the **CDL** (Central Dynamics Limited) PEC-102 system permits an economical configuration for a specific task along with an expansion capability to meet future requirements. The computer-controlled electronic-editing system requires only that each peripheral be capable of reproducing the SMPTE/EBU time code and controllable from electrical commands and that recording devices be equipped with switchable sequencers.

The software-controlled system performs the basic editor operations of synchronization, sequencing, storing, and switching. It also allows for interface units that adapt the specific peripheral equipment (quadruplex videotape, helical videotape, film reader, audio reproducer, or live studio equipment) to the control outputs. Tables of contents, maintained in the software, list the performance of each peripheral.

The first step of the CDL electronic-editing system, initialize (init), establishes machine assignments and operating parameters. Data is entered through a conventional keyboard with a display on a CRT terminal arranged for high update speeds. The system can control up to eight machines and operate an associated audio/video switcher that allows the programmer to select cross points, transitions, and effects; set up transition rates; and initiate the effects in both audio and video sections. The system's 600-scene file capability makes possible a fully automatic program assembly.

The tape-load mode (tlod) allows the editor to establish the details of the production and the tapes to be used. Entry is from the data keyboard to a list-format display, as in the init mode. A program module is loaded in approximately 100 milliseconds and started automatically by the executive module, which initiates a number of routines along with the input/output instructions for the peripherals such as the disk, display, keyboard, and teletype. The tape-load mode is generally required to store production details including scene identifications and valid tape addresses. These addresses and production lengths are passed to the data tables for subsequent recall.

In the operate (oper) mode the editor-programmer sets up the details of the edit sequence to be performed. All edit decisions relevant to addresses, durations, sequences, transitions, and effects are entered from the data keyboard into a simple interactive flow diagram of the sequences. The generated flow diagram, in addition to providing an instant view of the edit sequences, updates to accommodate source selections, transitions, and effects. Also indicated are the addresses currently accessed for entry or modification. The display format for oper is also used for machine control, with which it is closely associated. Addresses may be entered from the keyboard, from tape address tracks, or from mass storage. All current-segment addresses are displayed and may be modified in either direction or replaced from the keyboard. Edit sequences on tape may be previewed individually and automatically to achieve the precision and visual effect desired. Sequence addresses can be stored for record or rehearse operation or, in the edit file, for future use. Command data for an associated audio/video switcher may be entered from the control panel. Data consists of video effects (keys, mattes), video transitions (cuts, wipes, fades, dissolves), and video-transition duration (from 4 to 299 frames). Audio effects or transitions follow the video information. The control software is arranged to update the display and make the necessary calculations and checks automatically as transitions are entered.

The machine-control (mctl) mode is used to run and monitor the equipment; to program previews, rehearsals, and edits; and to operate the audio/video switcher. The system employs the same format as operate (oper), displaying the current position of each machine, the operational mode (cue, review, rehearse, record), and any error messages. In this mode the editor operates the tape machines and switcher by reading previously stored data and translating it into current time-code and control-panel commands. If he wishes, the editor can control the operation manually by idling the mctl mode. In the machine-control mode the system recues each machine to an address derived from oper (entry address) and init (preroll time). The cueing routine is entered automatically at the beginning of preview, rehearse, and record. All of the system's units are synchronized to frame accuracy by comparing addresses with the master machine. The record machine can make edits as small as one frame if necessary. In most cases edits are performed only if the machines are correctly synchronized. A warning message halts the procedure if the machines are not operating in sync or if the framing is incorrect. To perform an open-ended edit (no scene-exit address is specified), the edit is terminated in an orderly manner with a manual command, and the exit address is stored in oper along with other pertinent data.

The edit-file system referred to as search (srch) allows the programmer to establish, review, or delete a file of edit sequences. The search mode covers all operations associated with the maintenance of a file (store) of up to 600 edit-scene records, the additions to machine control (mctl) to effect continuous rehearsal or editing from the file, and operations associated with reading a tape to the file or outputting the file contents to tape. In the search mode all scene records required for each segment recorded on the master tape can be instantly recalled for display on the command console. The record provides a full definition of the segment including scene number and type, record address at the beginning of the segment, source-tape number, entry address, exit address, transition, effect, and rate. The search mode employs a list-format display of the scene records, arranged in order of execution. The records may be inserted or logged into the file either from oper or in blocks from an external tape reader. Keys are provided on the display/command console for inserting, deleting, reviewing, or modifying any record in the file. If errors are encountered, the incorrect information is output to the printer and an error message is displayed on the console. The editor or programmer can then correct the error from the printer keyboard by retyping the information. Records are output from the file to oper as required in order to update the system at the end of each programmed edit. The system is arranged to run continuously without intervention until no recue is available, a nonloaded tape is found, or a stop point is encountered. Insertion of the correct data causes the

system to resume automatic operation. The program may be carried out manually one scene at a time; automatic operation is on a scene-sequential basis.

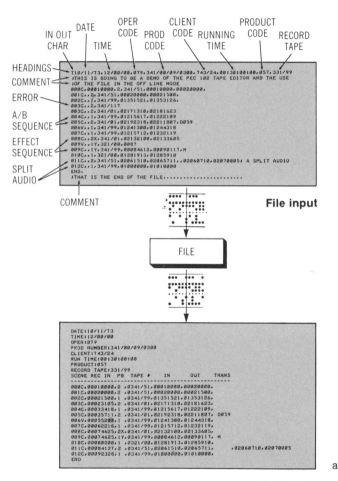

10-8. (a) Input-output format for PEC-102 edit file. (b) On-location and off-line editing applications. (Courtesy of Central Dynamics Ltd.)

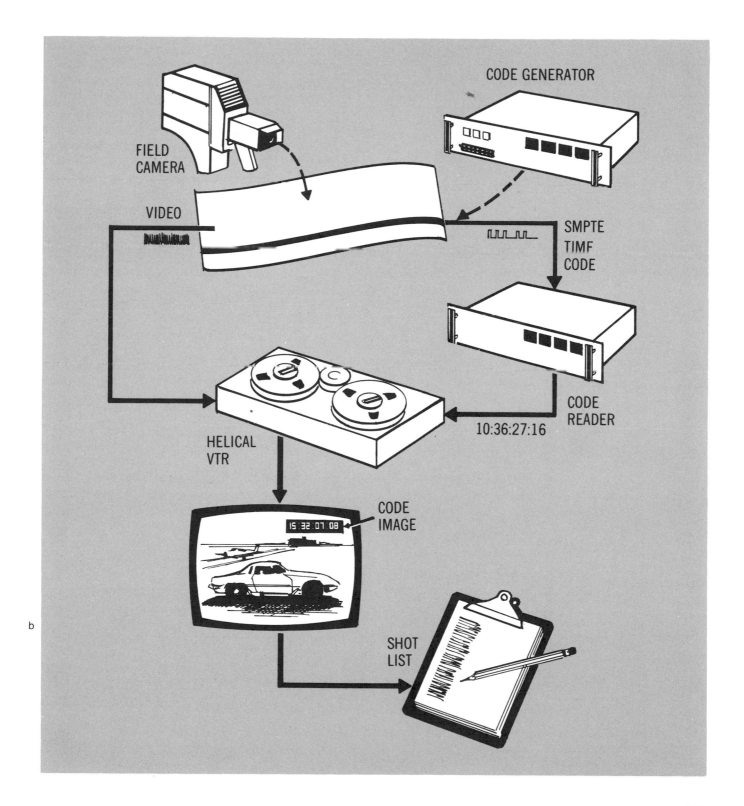

CODE GENERATOR

FIELD CAMERA

VIDEO

SMPTE TIMF CODE

CODE READER

10:36:27:16

HELICAL VTR

CODE IMAGE

15:32:07:08

SHOT LIST

b

155

Selection of the current operating modes rests with the editor-programmer, with some assistance from the real-time machine-control section. The system takes over once a mode is established. The Central Dynamics PEC-102 automatic-editor control brings the advantages of real automation to the videotape editing room in a most efficient, flexible format. The processing power of the digital computer is teamed with an optimized, interactive display system to allow the editor full artistic and content freedom, with all checks, calculations, and operations performed automatically and rapidly.

```
                   INITIALIZE

                ROLL     ADDR
RECORD           10        0
PLAYBACK 1       10        1
PLAYBACK 2       10        2
EXTERNAL         03        8




DATE: 10/11/73
TIME: 12:00:00
OPER: 079
```

The INITIALIZE Mode

```
              TAPE LOAD

PROD NUMBER: 341/00/09/0300
ERROR CHECK: 1            CLIENT: 743/24
TIME: 00:30:00:00         PRODUCT: 057


             TAPE NO.     FROM          TO
RECORD:       331/99    00:01:00:00   00:10:00:00
PLAYBACK 1:   341/99    00:08:00:00   01:37:00:00
PLAYBACK 2:   341/01    02:00:00:00   02:29:59:29


EXTERNAL :                :  :  :        :  :  :
```

The TAPE LOAD Mode

```
                OPERATE -  RECORD
     POS-REC         POS-PB1        POS-PB2
  00:03:24:18     01:21:46:17    02:19:13:18
  SCENE  004         COMB        00:26:05:05 TO GO

  00:03:34:18     00:03:57:01    00:05:51:20
  REC  <----------*---------------->

  01:21:56:17     01:22:21:09
  PB1  <------------->

                  02:19:23:18     02:21:18:07    D   059
  PB2             <---------------->

  EXT
```

The OPERATE and MACHINE CONTROL Mode   A-B Roll

↔
10-9. Operational modes and CRT displays. (Courtesy of Central Dynamics Ltd.)

CMX SYSTEM

Developed by the Columbia Broadcasting System and Memorex, the acronym **CMX** describes a computerized videotape editorial system. Available as either an on-line or an off-line computer-controlled video-editing technique, CMX represents a radical departure from the more conventional postproduction processes.

Designed for both tight budgets and big spenders, the CMX system is centered around a series of computer disk drives. These disk packs are random-access, unlike the reels of tape in conventional computer-storage systems, which must spin in order to reach a specific address or location. Stored information can be retrieved instantly without stop-and-go tape spinning and the accompanying search for identifying frames. The disks, loaded with buy scenes or selected takes, are capable of storing up to 25 minutes of material.

The stored picture information, although it may originally have been recorded in color, appears as black-and-white images on the system's monitors. The SMPTE time-code numbers, similar to edge numbers on motion-picture film, are encoded at the time of loading. These code numbers identify each frame in a scene stored on the disks.

The number of disks tied into a system varies, depending upon the particular system's configuration. Each additional disk increases the system's maximum storage capacity. Picture, sound, and time-code address information is loaded or transferred into the storage system for subsequent retrieval in the form of computer language. Each frame of elecronic picture information occupies a single band on the round disks. The time-code address that identifies each frame in a stored scene is the key to the retrieval process, in which any frame can be sliced from the disk at any time during editing for automatic assembly. If all or

```
           OPERATE - RECORD
    POS-REC        POS-PB1        POS-PB2
 00:07:36:12    00:08:36:12    02:13:11:09
 SCENE  008        COMB        00:23:54:25 TO GO

 00:07:46:25    00:08:02:00
 REC   <------------->

 00:08:46:12    00:09:01:17                    M
 PB1   <------------->

 02:13:21:00    02:13:36:05
 PB2   <------------->

 EXT
```

Simultaneous Roll

```
 FETCH:004            SEARCH
    SCENE REC IN  PB   TAPE #      IN       OUT      TRANS
    002C 00021500  1  0341/99  01351521 01353126
    003C 00023105  2  0341/01  02171310 02101623
 >>004C 00033418  1  0341/99  01215617 01222109
    005C 00035711  2  0341/01  02192318 02211007   0059
    006V 00055200  1  0341/99  01241300 01244318
    007C 00062210  1  0341/99  01215712 01232119
    008C 00074625 2X  0341/01  02132108 02133605
    009C 00074625 1Y  0341/99  00084612 00090117   M
    010C 00083200  1  0321/00  01281913 01295910
    011C 00084127  2  0341/51  02061510 02065711
    011A 00084127  2  0341/51  02060710 02070005
    012C 00092328  1  0341/99  01000000 01010000
    013END
```

The SEARCH Mode

any part of the material to be used during assembly is on motion-picture film, that length of film must be transferred to tape. The original production tapes, after being transferred to video disks, are stored in areas in which temperature and humidity are precisely controlled.

Scenes stored on the disk may or may not be in numerical order. During the editing process scenes are not moved from one part of the storage disk to another: the immediate availability of any scene or frame within a scene in real time negates the need for any specific or prescribed order for loading and storing picture information. Even as the editing process progresses, the outgoing scene is not repositioned to precede the incoming scene on the storage disk. The punched control tape containing the editor's decisions recalls any scene from any part of the disk pack at the required point during the assembly procedure. The time-code information added initially assures complete frame-count accuracy during assembly and precise arrangement of scenes in a stated sequence.

Seated at the control console in front of two monitors, the director and editor begin the post-production process by looking at the right-hand monitor and reviewing the displayed menu. All of the previously loaded scenes, now under computer control, are identified by a word or number. The editor or console operator, following the director's instructions, touches one of the scenes listed on the menu with a small, pen-shaped flashlight. The first frame of the selected scene instantly becomes visible on one of the console's monitors. A list of control functions (play, splice, dissolve) are displayed simultaneously. The scene may be viewed at any convenient speed selected by the director. The first frame of the opening scene is transferred to the monitor that is used for the actual assembly of the scenes stored on the disk. During the editing process the scene being edited is viewed on the left-hand monitor; the menu is displayed on the right-hand monitor, on which entry decisions are also made. The first frame of any scene selected from the menu by the light pen is visible on the right-hand monitor, along with the possible options such as normal speed, fast forward, reverse, or single frame.

During the editing process the programmer at the console uses the light pen to retrieve scenes stored on the disks or to issue commands based on the items listed on the menu displayed on the right-hand monitor. The mere touch of the light pen on a specific command listed in the display activates the entire system and transmits edit decisions that can subsequently be reviewed in real time. The pencil-sized electronic device in the programmer's hands is not only used for scene selection and decision making but also for controlling the mode of operation. If desirable, the scene can be viewed on a frame-by-frame, stop-motion basis, with the action frozen at any specific frame. The programmer at the control console allows the scene to run to the frame at which it is to be cut or to which a transitional effect is to be added. At that point the programmer points the light pen to another specific item on the menu displayed on the right-hand monitor. The director can see the first few frames of the incoming scene as well as the last frames from the outgoing scene. If the cut is satisfactory, the next scene is reviewed and the editorial process continues.

The editor and programmer are seated at the control console and can review both the entry and exit points of the outgoing and incoming scenes on the dual display monitors simultaneously, so they can assess the editing decisions before going on to the next scene. If the transitional between-scenes effect initially programmed does not produce the desired visual result, a new effect can be programmed immediately. The light pen encourages experimentation and the use of enhancing optical effects. Some of these effects could be produced optically on film only with a considerable amount of time and effort. The electronic, computer-generated optical effects, by comparison, can be programmed and previewed before the outgoing and incoming scenes are married to everyone's satisfaction. In addition to transitional effects superimpositions can be programmed for insertion over any part of the scene. Lengthening or shortening actions within scenes is yet another option open to the editor. All editing decisions and assembly instructions are noted in the computer processor. Entered in the computer memory, the resultant punched control tape assures precise fulfillment of all programmed commands during the assembly process. After the last cut is made and the composite results are previewed, the director, editor, and programmer know exactly what the scenes look like. There are no dailies to screen and no unanticipated surprises. Executive approval initiates the assembly or conformation process.

The assembly process begins when the punched control tape is programmed back into the system's memory. Quadruplex recorders carry the original reels of recorded picture and sound information. The assembly process is completely automatic. The computer synchronizes all the recorders tied into the system and transfers or rerecords the selected scenes along with the computer-generated special and transitional effects and superimpositions to produce a splice-free, composite edit master that is complete with respect to visual continuity. The edit master can be produced at any time after the edit decisions have been made and the instructions filed away in the form of the punched control tapes. The director and/or editor can leave the assembly task to the programmer, who can complete the postproduction chore at any subsequent time when the videotape-recording equipment is not in use. Unlike the film-assembly process, in which each edit change or revision is accompanied by fresh splices and the jogging of successive frames of film through the moviola's viewing head, the videotape editor does not physically handle the tape to adjust sync or change cut points. The videotape editor or console operator types the changes in computer language on the typewriter's operating keyboard and substitutes the new section of control tape for the tape that was previously punched. The composite result on the 2"-wide rolls of videotape, referred to as the edit master, is analogous to the optical negative in motion-picture production. Produced in on-line systems by the control tapes, the edit master is complete with respect to scene lengths, special and transitional effects, and superimpositions. The edited and assembled rolls may be duplicated and distributed as videotape or reproduced on videocassettes. Newer CMX systems—the sys-

tem 40, for example—are designed to make electronic news gathering (ENG) faster and more efficient. Appropriately identified as the **news editor,** the system consists of a computer, related CMX software, operating keyboard, interconnect cabling, and several Sony U-Matic videocassette recorders.

Electronic-editing systems are having a more or less anticipated effect upon both motion-picture and videotape production techniques. The ability to see a sequence of action, as on motion-picture film, is no longer limited to projecting the processed lengths of film on beaded screens or moviola viewing heads. The images recorded on the 2"-wide rolls of videotape can be viewed on monitors without chemical processing and while the performers are still speaking their lines. The playback can be reviewed at normal speed, slowed down, speeded up, reversed, or jogged for single-frame, stop-motion analysis. Computer-controlled systems provide the producer with a number of options. The subject matter can be photographed with electronic cameras, edited with an electronic system, assembled on videotape, and telecast in that form. As an alternative the subject matter may be exposed on film with conventional motion-picture cameras and transferred to videotape for editing, assembly, and broadcast. There are many variations—the production process may begin with film, be subsequently transferred to videotape for editing and assembly, and be transferred back to film for theatrical release or television broadcasting.

A comparison of the relative economics of conventional and electronic editing and assembly processes requires qualification: unusual circumstances would negate any arguments offered on behalf of either system. The cost of editing and assembly cannot and should not be considered as a completely separate budgetary item. Film production in conventional film systems is comparatively inexpensive, but raw stock, processing, and opticals tilt the scales in the opposite direction. Electronic systems initially require much higher capital outlays, but consumables and operation are considerably cheaper.

Using single-camera techniques and shooting in random sequence, the director and editor can approach the editing and assembly process with the knowledge that each scene that is transferred automatically into the computer storage can be retrieved instantly from the disk-storage system. At the control console the director and/or editor can view outgoing and incoming scenes simultaneously on dual monitors and check the exit and entry points. The editor can see the actual cuts without making a single splice; special and transitional effects can be viewed without producing costly opticals; the producer can rely on the fact that the punched control tape automatically fulfills every editing and assembly instruction and produces a composite product that is complete with respect to picture and sound continuity.

SONY U-MATIC RECORDING/EDITING SYSTEM

Sony U-Matic 3/4" videocassette recorder (model VO-2850) with its electronic-editing capability fills the need for a teleproduction video-editing machine. The unit combines the advantages of a high-quality production recorder with the ease of operation and speed of an automatic postproduction editing machine. The 3/4" tape format, with its book-size cassette, makes distribution of fully edited tapes a quick, simple, and economical process.

With the U-Matic videocassette recorder the problems associated with threading and handling tape are nonexistent. Tapes may be stopped at any point and removed without rewinding. The advanced design of the tape-transport system and the new solenoid controls offer ease of operation and reliable performance over extended working periods. The unit also incorporates a memory of commands—logic circuits memorize the sequence of selected tape functions, reducing actual operating time.

The 3/4" tape passes over the recording head at a speed of 3 3/4" per second—considerably more economical than the 2"-wide tapes that are recorded at 15" per second. Despite the low tape speed the quality of the signal is comparable to professional broadcast standards. Operated from any 120-volt alternating-current power outlet, the teleproduction video-editing machine incorporates technology that provides flexibility in both production and postproduction requirements. Insert-and-assemble editing is accomplished with the use of a flying erase head and occurs in the vertical interval of the picture. This feature assures quick, professional edits and a signal-to-noise ratio of 45dB. The unit has two high-fidelity audio tracks that may be independently edited with or without video. The high-speed decay circuits assure smooth audio-edit transitions without the usual objectionable pops.

COMPUTER-CONTROLLED PROCESSING AND PROJECTION

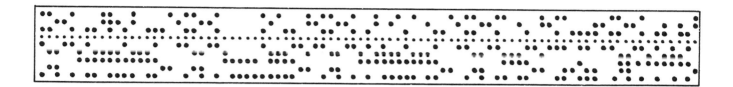

The earlier of the two methods generally used for producing an image for subsequent projection is the conventional system, in which lengths of sprocketed film are exposed by means of a motion-picture camera. The latent images on the negative film are made visible by chemical processing, and the system involves the use of master positives, dupe negatives, optical negatives, and release prints. The newer electronic system, by comparison, utilizes the television camera and records the images on videotape. The need for chemical processing is eliminated, since the lengths of videotape are available for playback at any time after the recording process is completed. Image converters are used to produce a dupe negative if the tape is transferred to film; this negative in turn is used to produce the release print.

The newer processing equipment is matching strides with other computer-controlled motion-picture-production processes. Representative of the newer electronic systems, for example, is the IBM System 7 computer for evaluating and monitoring the operation of the light valve in the Bell and Howell additive-color printers. This system earned a Class II Academy Award for Consolidated Film Industries. In this computer-controlled processing system the computer notes the work-order number, film-reel number, negative emulsion type, and emulsion number. Also stored in the computer memory is the printing-machine number, time and date of printing, and film-footage count.

CFI's automatic printing process begins when the printer's start button is depressed. The printing lamps make the adjustment from stabilized stand-by voltage to printing voltage and amperage. The punched control tape verifies the sensitometric setting for the raw-stock emulsion threaded into the processing unit. The machine starts automatically as soon as the setup is validated. Any error in the system illuminates an error light that indicates the reason for the malfunction. The warning light deactivates the control mechanism, and the printer is immediately disabled. In addition to the central computer devices attached to each printing machine (indicating the printing-lamp voltages and amperages) act as an interface between the printing machine and the computer. Bell and Howell's newer automatic additive-color film printers feature photoelectric reader heads and incorporate digital frame-count cue displays of the printer position. The printer's electronic integrated-circuit-type design is ideally suited for computer-controlled operation.

Technicolor's computer-controlled film-processing system, in line with the current trend, uses punched tape generated by the typewriter address. In the Technicolor system the computer has international input-power capability in that printer light timings originally used for processing material in the company's overseas plants are converted into B & H light equivalents. The new timings are made to conform by revising existing tapes or by feeding the scene-to-scene light information into the computer by means of a typewriter address to produce new control tapes. The computer is also able to provide hard-copy readouts. Changes in the timing of negative material are made either as overall corrections or on a scene-to-scene basis.

The Hazeltine film analyzer provides timing data for negatives, internegatives, and reversal film. The analyzer's electronic control panel includes digital readouts for each of the primary colors. Individual gamma controls make it possible to match any of the primary colors in the negative to the print film's gamma characteristics. Except for the system's flying-spot scanner and photomultipliers the analyzer is a solid-state unit. A related system developed for use in conjunction with the color-film analyzer is the time-card-preparation system. Three groups of pushbuttons offer a total of 125 possible combinations and provide hard-copy printouts identifying image and scene-content description, including items such as subject, location, and camera angle. Scene numbers, special effects, and printer-light data for each of the primary colors are also listed in the printout. Frame-count cueing and exposure information is entered by means of a decimal keyboard.

Complementing the newer electronic imaging techniques are the computer-controlled projection systems. The United Audio-Visual Corporation's Multivision system, for example, can control a number of projection units and keep them synchronized with each other during the entire program. Each of the control units tied into the system contains a computer-punched tape reader, silent high-speed stepping motor, and solid-state memory store. Plug-in printed circuit cards are used throughout for all control circuitry. The memory programmer automatically stores information ahead of the program so that all of the required control channels are activated instantaneously by the audio cue. The system includes sound replay,

memory programmer, and projection-dissolve control units. The Cuemaster multimedia program console provides the means for controlling movie projectors, slide projectors, tape machines, dissolve units, stage lighting, stereo sound, animated displays, and/or any electromechanical devices tied into the system. Designed as both a programmer and an electronic keyboard control for the tape punch, the Cuemaster's program signals can be transmitted over telephone lines for precise control in performing simultaneous shows in remote cities. The UAV Screenmaster module represents an impressive technological advance in multiscreen techniques. Capable of producing superimposition, composite, and/or overlay effects, the individual programmable command functions include quick cuts, dissolves (from 2 to 15 seconds), and freezes on any of the screens in use. The Screenmaster also drops the house lights to half wattage, turns them off, or switches them back to full brightness, depending upon the information cue fed to the control tape during the programming process.

The 1″ tape that controls the entire projection procedure has 8 data-hole positions across its width. Each hole is assigned a number from 1 through 8, and each single line carries a complete, specific cue. The numbered squares in the main panel are actually pushbuttons that refer to both the data-hole positions on the tape and to specific projector channels. Manually pressing any channel button arms that particular circuit and causes the bulb concealed under the button to light: the lit circuits are sequence-locked for a single cue, but nothing happens until the screen action is cued or the tape is punched. In this way each cue is committed to memory on the punched tape. Working against a prepared cue sheet and with the tape running in real time, the programmer sets the cue on the panel button and at the appropriate moment presses the button to record the cue. With the cue-mode switch set to real time the audio Cuemaster receives the digital signals from the completed tape during playback and causes the appropriate screen actions to occur in synchronization with the previously programmed instructions. As each cue passes through the Cuemaster, the numbered panel pushbuttons flicker momentarily. Each of the prerecorded cue signals may be confirmed during an optional dry run or rehearsal.

The operator can hold a cue rehearsal and confirm the programmed instructions without connecting the projectors. As the finished program is played back on the reader head, each cue on the punched tape lights channel numbers corresponding to the holes punched in the tape. In either mode the lit channels provide a preview of the upcoming

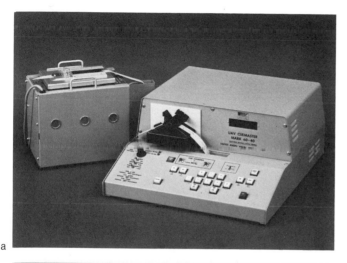

a

b

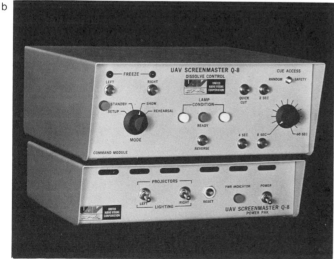

11-1. (a) The Cuemaster with a control-tape punch. (b) The Cuemaster control panel.

screen action. Since the operator is able to monitor the program and anticipate the next tape cue, he has access to all channels and can correct, skip, or add a cue at any point—either during the rehearsal or during the actual screening. Corrections to the punched tape are made without splicing or correction tabs. When the rehearsal is completed, the punched tape is threaded in the reader head, and the cue counter is reset to zero. Since every line of the punched tape has an assigned counter number, cues are readily identifiable. If the rehearsal is satisfactory, the electric punch duplicates all the good cues at the rate of 20 lines of tape per second. If a programming change is indicated during the tape-duplicating process, the programmer merely switches the cue speed to normal and enters the new instructions manually one tape line at a time. Operable by any momentary-closure device, the forward button activates the Cuemaster. A panel indicator light is pulsed by any remote cueing device for visual proof of the equipment's performance. The panel light also indicates any malfunction of the remote cue device.

Representative of electronically controlled projection systems is the RCA FR-35 servo-controlled 35mm film projector. Designed for telecine use, the projector includes a number of features such as instant start and stop, variable speed controls, and individual frame cueing. Operating in either the forward or the reverse mode and at normal or high-speed rates, the system provides extremely accurate picture registration. Used in conjunction with a TK-28 color-film pickup camera, the FR-35 demonstrates its versatility in the telecine chain. The TK-28, equipped to correct errors in a telecine system, controls black-and-white levels, flare, black-and-white balance, and gamma. It can also be used with the TCP-1624 television cartridge-film projector. The TCP system includes two self-threading 16mm projectors that accept film cartridges with running times from 10 seconds to 2 minutes. In this film-handling system as many as 24 cartridges can be loaded on removable magazine trays. Ideally suited for automating the playback of film commercials on television, the TCP-1624 can be used for either single-play or automatic-sequence operation.

The raster-bending techniques described in these pages represent impressive technological advances in the field of electronic imaging. To say that the new, mind-boggling means of communication literally materialized rather than developed would be a gross understatement. The infant industry began taking giant strides without ever having experienced the coddling comforts that accompany the crawling stage. Given impetus by related industries and cooperative equipment manufacturers, the gyrating graphics made possible by electronic imaging techniques have gained immediate and wide acceptance as a uniquely different art form. Television's commercial break, with its analog animation, leaping logos, and twisting titles, has been given new meaning and a new look. The eye-arresting, illusory electronic effects incorporated in many of the more recent 1-minute featurettes have kept viewers interested, products moving, and sponsors happy. The technique has so much potential and offers such great promise in so many areas that any prediction on the direction that the new art form might take can be nothing more than mere conjecture. The future will almost certainly be determined by the imaginative directors and producers whose talents include the ability to push buttons, flip switches, turn knobs, and spin dials on manually manipulated mechanisms such as video synthesizers.

COMPUTER GLOSSARY

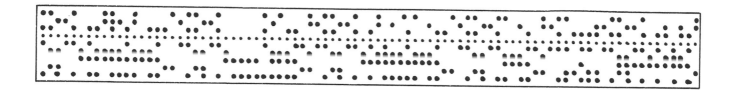

A & B composite system   See **intersync system.**

A & B editing, videotape   The preparation of material recorded on videotape and subsequent production of a composite tape that is complete with respect to visual continuity. The A and B rolls used for the video mix are prepared in the same way as for a ·16mm motion-picture production. During the assembly process each edited scene is added sequentially along with transitional and special effects in building-block fashion. The picture and sound information, carried on separate rolls of videotape, are combined on the composite tape.

abscissa   A programming entry or reference input that defines the specific position of the x-coordinate of a point in the display.

access time   The interval between ordering and delivery of information from a retrieval system. **Instant access** denotes a retrieval time of such short duration that the eye is unable to recognize a time lag.

address   An identifying code or label that specifies the location of stored information in a retrieval system. In videotape recording the identification is usually made by a decoder, which displays a cue track's serialized address on a nixie tube.

aerial   See antenna.

Algol   An acronym for algorithmic language, a high-level computer-programming language.

algorithm   A rule of procedure for solving a mathematical problem that frequently involves repeating a particular operation.

alphamerics   A set consisting of both the letters of the alphabet and numbers from 0 through 9, which are programmed via a teletype and used in a display to provide captions and/or information relative to the picture over which they are superimposed. Alphameric characters are also used in some computer-generated imaging processes—Tarps, for example—to produce gray levels.Different shades and brightness values of gray are produced by overstriking or superimposing several letters, numbers, or symbols in the same space.

amplifier   A device that increases or reinforces the strength, current, or voltage of a signal or an electric impulse.

Amtec An acronym for automatic time-element compensator, an Ampex accessory that eliminates skewing and quadrature error during playback.

analog computer A computer system that differentiates and integrates the continuous input of image-forming elements on the basis of changes in the magnitude of the signals. Unlike the digital computer, in which all the elements are similar, the analog computer contains a grouping of dissimilar elements. Each of these electronic circuits can resolve an individual problem in a complex command—and all operate simultaneously. Because of the dissimilarity of the circuits the computer's response is not always a positive "yes" or "no" but occasionally a "maybe" if appropriate. Any answer is subject to change with the introduction and programming of new or additional information.

Animac A pioneer analog-computer system for generating movement with static graphics.

antenna, or **aerial** A skeletal-steel structure, metal rod, or reflecting disk used to transmit and receive sound and picture information in the form of electromagnetic waves and other signals.

arc One of the drafting routines in computer programming: call arc = $(x_1, y_1, x_c, y_c, \frac{cw}{ccw}, n, \text{width}, x\text{-limit}, y\text{-limit}, \text{frames}, \text{speed})$ describes an arc beginning at (x_1, y_1), with center at (x_c, y_c), and proceeding either clockwise (cw) or counterclockwise (ccw). The arc stops when the line reaches a specific x- or y-limit or when a complete circle is drawn.

are The abbreviation for area in graphic-oriented routine-initialization codes; it specifies the vertex description of a geometric area.

array A grouping or series of picture elements in a raster-scan representation.

audio A generic term that describes sound in recording, transmission, or reception operations.

audio-cue channel The voice tones or digital cues recorded along the edge of a length of videotape. These signals have no effect on any previously recorded information.

audio head The device in a tape-transport mechanism over which the tape passes to record the sound. It encodes the sound signal over the tape's iron-oxide layer. The sound signal precedes the corresponding picture by 9 1/4" on videotape.

Autocharts A feature of the Genigraphics system designed to satisfy recurring requirements and update specialized charts. The key to the system is a slide catalog of prepared graphic formats, number-coded in the computer's memory for easy access and individual specifications.

bac The abbreviation for background in graphics-oriented initialization codes.

banding A disturbance in videotape-recording playback, characterized by color variations in separate horizontal segments and caused by small differences in head speed between taping and playback recorders. An automatic velocity compensator corrects this visually annoying disturbance.

bandwidth A specific frequency range between high and low parameters, measured in cycles per second.

bank A number of related controls grouped in a specific section of a computer-programming, videotape-recording, or television-transmission console.

bcd An acronym for binary-code decimal, a numbering system in which 0 and 1 are the only symbols used to represent decimal numbers.

Beflix A computer-programming language used to perform drafting-type operations that are logically simple to describe but graphically or temporally complex in execution.

bgread A miscellaneous operation in the Fortran IV Beflix programming language that updates the position of the image-forming elements on the grid after a subroutine.

B-H curve Analogous to the characteristic curves of different film emulsions, a B-H curve indicates the characteristics of a given type of magnetic tape. The plottings include information on output levels, degree of noise, and distortion levels.

binary system A numbering system that utilizes only two symbols, 0 and 1, to indicate a positive condition, such as on or off and/or yes or no.

bit An acronym for binary digit, a unit of information indicated by either 0 or 1.

bit rate The rate at which a bit travels between two specific points or passes a specific point.

black level (1) The darkest portion of a transmitted picture. (2) The minimum signal strength needed to transmit an image.

blanking The part of the scanning process in which the scanning beam, after completing its sweep of the target area, returns to the top of the frame to begin the procedure for the next picture. The process is repeated 30 times per second. See also **field** and **frame.**

booster signal See **microwave link.**

breakup A momentary distortion of an image.

broadcast band, or **broadcast channel** The bandwidth, or range of frequencies, assigned to a station for transmitting program material. The bands are assigned by the Federal Communications Commission (FCC).

bug A base pointer or scanner that symbolizes the movement of a picture element from one coordinate position in the grid to another. It moves around the surface of a two-dimensional grid, reading and/or changing the number that it occupies.

A bug is identified by one of the letters of the alphabet: a through m are used for reading and writing the grid; n through z, for moving or retaining programmed positions. Before a bug is used, it must be initialized or identified at its coordinates by the signal call place (bug, x, y). It can be moved to a new x- or y-coordinate with the routine call move (bug, to, bug 2).

byte A square adjacent to specific x- and y-coordinates, usually on the same horizontal grid row. It consists of a grouping of eight adjacent bits and is processed as a unit.

Caesar An acronym for computer-animated episodes using single-axis rotation, an analog-type computer-animation system.

camera A miscellaneous operation in the Fortran IV Beflix programming language that controls output. The parameters for the routine are call camera (frames). If the installation provides for various options, the parameters must be preset by other subroutines.

capture mode An operational mode that provides the graphic artist with cursor-selected control of any component part of the artwork.

cartridge A container of tape or film, fed from a single reel in a continuous loop. Compare **cassette.**

cassette A container in which the film or tape is wound from the feed reel to the take-up reel. Compare **cartridge.**

cathode-ray tube (CRT) An electron tube in which electrons from the cathode are directed at a target area (the fluorescent screen) at the end of the tube. The fluorescent screen glows wherever the electrons strike in direct proportion to the beam's intensity (the magnitude of the input signal).

center A rectangular-subarea operation in the Fortran IV Beflix language that controls the centering of typed captions or labels within a specified rectangular subarea in the grid. The parameters for the routine are call center (right, top, left, bottom).

channel See **broadcast band.**

Charactron tube A device in computers such as the Stromberg Datagraphics 4060 microfilm printer that provides a number of distinguishable characters used to produce electronic images. Both the original Beflix system and the Tarps language produce tonal values and textures with Charactron symbols.

chroma The color intensity, or saturation, of a picture.

chroma-key process A matteing system used to insert picture information from one source into a specific area of a second picture. Unlike the high-contrast mattes used in motion-picture production, the chroma-key process uses controlled colors (usually blue), and, unlike the optical processes needed to produce the composite effect in film, the results produced with this electronic keying system are immediate and can be monitored or previewed before the actual taping begins.

cinching Analogous to the buckling of film in a motion-picture camera, the folding over of videotape layers to produce an accordionlike appearance. The effect is caused by the sudden stoppage of the transport mechanism that controls the movement of the tape from the feed reel past the record or playback heads to the take-up reel.

cine rate A picture-display speed that is compatible with standard motion-picture photographic and projection rates—24 frames per second.

Cinetron An on-line digital-computer system for recording animation and opticals.

cir The abbreviation for circle in graphics-oriented routine-initialization codes. It is used to specify the size, color, and position in the frame of the circle.

clipping The regulation of a videotape signal by removing extreme high and low frequencies in order to keep the signal within preset limitations. The process produces a more favorable balance by reducing the range between intense highlights and dense shadows.

closed circuit (CC) The transmission of picture and sound information to audiences at specific locations by means of coaxial cables. The transmitted signal is available only to outlets connected directly to the cables.

clr An abbreviation for color in graphics-oriented routine-initialization codes, used to select color schemes.

coaxial cables A cable used to transmit broadcast material directly to specific locations. The broadcast information is transmitted at high frequencies with a minimal loss of signal strength.

Cobol An acronym for common business-oriented language, a standardized language for computer programming.

code generator A coding device that puts an identifying label on a videotape that carries picture and/or sound information. See **address.**

coding The process of recording an audio code on the videotape's cue track. Similar to the film-edge numbering system, each frame on the reel carries its own identifying number.

color balance The process of adjusting the color output to match the standard color bars that appear at the beginning of each tape. If several cameras are used for taping, skin tones are used as the balancing factor.

color bars The color standards of the Society of Motion Picture and Television Engineers (SMPTE), which appear at the beginning of each tape and ensure uniform color balance in material used for broadcast purposes.

color mode An operational mode with which the color of selected objects can be regulated through independent hue, chroma, and value controls.

Colorpletor An encoding module that combines the red, green, and blue color components in the video signal to produce a single transmission signal. The colors in the composite signal are restored or duplicated within the television receiver by means of a decoder, reproducing the color values in the original subject matter.

combin A Fortran IV Beflix rectangular-subarea operation, used to program the combination of picture elements from two grid areas. The parameters are call combin (right, top, left, bottom, onto-right, onto-top,
  big
  small).
  sum
  diff

compiler A computer routine that automatically produces a program for a particular problem.

computer See **analog computer** and **digital computer.**

computer-output microfilm (COM) A system in which each computer-generated frame presented to a high-resolution CRT is recorded on microfilm. The microfilm camera, slaved to the computer, provides a convenient method for permanently recording graphic output.

continuously variable display images, dynamic computer-generated animation imaging, or **dynamic computer imaging.** A method of generating and recording graphics, based on the analog computer, in which images can be displayed and altered while they are being created.

contrast ratio The brightness range between the deepest black and the whitest white in a reproduced picture.

control panel The control and monitoring equipment for recording, reproducing, and transmitting broadcast material.

coordinates In computer terminology, the lines or reference points that are used to define specific points within a stated area. A lowercase x describes a horizontal reference point; y indicates a vertical point; z indicates dimension.

copy A Fortran IV Beflix rectangular-subarea operation, used to copy an area, specified by the first four parameters, in a second location, the top right-hand corner of which is defined by onto-right, onto-top. The parameters for the routine are call copy (right, top, left, bottom, onto-right, onto-top).

CPL An acronym for circular plot area in data-plotting initialization codes. It describes a pre-programmed plot area for pie charts.

CPU The abbreviation for central processing unit, which includes two units, arithmetic and control, that combine to operate on data.

CRT The abbreviation for cathode-ray tube.

cue In computer programming and videotape production an identifying code or label or a reference point. The cue track in a videotape recording contains reference pulses or time codes that are used by the editor in the postproduction stages.

cue track An area that runs longitudinally along the videotape and carries audio and editing information in the form of reference pulses or time codes.

cursor An instrument such as a mouse or a light pencil that is used to generate graphics in the computer or to recall stored scenes during the editing process.

cutoff area, or **safety area** The part of the film frame that is not visible on monitors or receivers during transmission. It is clearly marked in field guides and camera viewfinders to indicate the position of titles and other superimposed material.

data Information or material to be computed. It may be in the form of perforated tape, as in off-line or on-line systems; punched cards, in the Beflix and SynthaVision systems; keyboard operations; or an external source, transmitted by means of a transducer. Data-input signals are translated into digital form for processing.

data camera A three-piece instrument that can be used with a variety of standard 1″ pickup tubes. The CVI 502 model accommodates standard vidicons of either separate or integral mesh, silicon-diode target tubes, lead-oxide target tubes, and other devices with appropriate mechanical and electrical characteristics. The camera is primarily intended for laboratory and industrial use and incorporates a number of features not normally found in closed-circuit-television (CCTV) equipment, such as externally controllable sweep circuits that can match nonconventional scanning patterns and rapid changes in position, angle, or size. Horizontal and vertical deflection may be interchanged; externally controllable beam blanking is provided for target-integration or pulsed-light applications; wide-range sweep circuits (DC-coupled) and video circuits are capable of operating at rates from 1 frame per second to 1,000 frames per second. Options include video sampler, shading generator, and digital control of various functions. The camera has a three-chassis configuration consisting of head, control unit, and power supply.

Datavision A video character generator developed by the 3M Company, which features internal video mixing and a built-in audio-recording storage interface.

debug (1) A corrective measure to put a malfunctioning piece of equipment back in operation. (2) A miscellaneous operation in the Fortran IV Beflix programming language that prints out a subarea and lists all of the bug positions within that area. The parameters for the routine are called debug (right, top, left, bottom, width, height).

decibel (dB) A unit of electric power used for measuring the volume of recorded sound.

decode In computer terminology, the process of translating computer-generated electronic equivalents back into the original picture-information components.

density analysis See **slicing.**

detail In computer graphics, the quality of the processed image as compared with the original input. Quality is judged by the number of bits, or picture elements, that are reproduced in the display.

digital computer A computer that consists of a single class of circuit elements, which are either on or off, or capable of responding to a programmed command with a yes or no answer. The digital computer is arithmetic and logical. Computations are made in sequence on a step-by-step, building-block basis. The input signals used to form an image must be translated into numerical equivalents and programmed sequentially. Complex problems may be resolved by interconnecting a number of electronic circuits, and each phase of a multifaceted problem is processed in turn. Digital image-generating computers are ideally suited for the complex mathematics needed to produce, program, and display the movement of figures, objects, shapes, and abstract designs.

direct operation See **on-line operation.**

disk recording A magnetic recording on a disk similar to a long-play record rather than on video-tape. Disk recording makes possible a variety of effects such as slowing down, speeding up, or freezing a sequence of action. The technique is used especially for instant replays during sports telecasts. See **videodisk.**

distortion A change in the signal wave form, compared with the original input, during transmission or amplification.

documentation A detailed record of programming instructions, valuable if and when subsequent modifications are required.

double-system recording In videotape and film, the simultaneous recording of picture and sound as separate elements. In addition to providing the editor with a greater degree of flexibility the overall quality of the recorded sound track is greatly improved. Compare **single system recording.**

downstream On a length of videotape a specific location some distance from a stated reference point in the same direction in which the tape is traveling.

downtime A period of time during which studio equipment is inoperative, such as for repairs or maintenance.

drafting operations Computer drawing of rectangles, straight lines, arcs, curves, and alphameric characters. All of these drafting operations (rect, line, arc, trace, and type) are dynamic in that they contain calls to camera, the output subroutine, with the exception of the rectangle-drawing operation.

dropout The loss of picture information due to physical defects in the videotape, particularly missing iron-oxide particles on the recording surface.

dropout compensator A sensing device that compensates for losses in the recorded picture by inserting matching picture elements in areas that need correction.

dtx The abbreviation for default text in graphics-oriented routine-initialization codes. The routine is used to indicate the preprogrammed size, position, and color of the text.

dub (1) or **dupe** The duplication of a videotape recording. (2) A mixing process in which a number of sound tracks (voice, music, and sound effects, for example) are combined to form the composite track. (3) A rerecording process in which new dialogue in a foreign language is substituted for the original dialogue.

dump A slang expression indicating dissatisfaction with the work produced to date.

dynamic In computer programming, a description of any subroutine that contains a call to the output routine. See **drafting operations.**

dynamic computer-generated animation imaging See **continuously variable display images.**

dynamic computer imaging See **continuously variable display images.**

ebr The abbreviation for electron-beam recording.

edgel A picture element that marks a boundary or forms a border between two adjacent rectangular subareas within the grid.

Edicomp An electronic-editing technique in which editing data is used to transfer the original sound and videotape to an edited second-generation copy.

Editec An electronic system in which the edited work print is used as a guide for rerecording the composite videotape.

edit sync Audible cues recorded in the videotape's cue track in the form of beep signals. Spaced at 1-second intervals, they help the editor to locate specific frames in the recorded videotape reel.

ee master An electronically edited tape produced from the original or master tape.

electron A negative electrical charge, many of which are released by the cathode in an electron tube.

electronic camera A recording device analogous to a motion-picture camera except that the images are not exposed on film: they are received as light energy, and the component parts are converted into electric impulses. These pulses or input signals can be displayed immediately or recorded on videotape for future use.

electronic editing A postproduction process in which picture and/or sound takes are selected and assembled electronically to produce the final composite print. The predominant time code is 1/30 second.

electrovisual systems Electronic-animation systems such as Animac, Caesar, and Scanimate are used extensively for producing films and videotapes for advertising, education, and scientific research. Animation of black-and-white line drawings into moving, contorting, full-color graphics and cartoons is seen frequently on commercial and educational television. There are many examples demonstrating an extensive capability to produce animated video displays, either of simulated object movement or data organized and treated in a wide variety of ways. The technology underlying representative systems includes digital and analog circuits as well as a complete video signal-handling and processing capability. Analog circuits embrace high-speed computation circuits employing summing amplifiers, inverters, multipliers, dividers, integrators, differentiators, and logarithmic amplifiers in addition to analog track-and-hold circuits, commutators, analog-to-digital and digital-to-analog converters, function generators, spectral-analysis filters, and phase-locked loops for coherent signal detection. Likewise, digital circuitry designs include numerical-computation circuits, time-code generators, parity generators, error correctors, magnetic tape-deck controllers, and circuits to interface with digital computers. The animation computers also employ digital timing and control logic. Electrooptical equipment includes high-resolution (x, y) television cameras employing vidicon, plumbicon, and silicon-diode pickup tubes; high-resolution CRT monitors; flying-spot scanners; scan-conversion equipment; and gray-scale color-encoding units. The latter provide an overall system resolution of 700 to 800 lines, as measured on a standard TV resolution chart. Circuit designs include wide-bandwidth video amplifiers, video threshold detectors, video switchers, and CRT deflection amplifiers.

encode To use a programming language, recording device, or imaging generator for translating picture information into electronic equivalents.

erase head One of three heads located in the tape recorder's transport mechanism, used to clean the tape for maximum signal-to-noise ratio.

error rate The number of mistakes, due to human error or mechanical failure, in relation to the total amount of information fed into a computer by the programmer.

evaluation A review or determination of the quality of the reproduced picture information.

expand A rectangular-subarea operation in the Fortran IV Beflix programming language. It expands the previously programmed picture, centered at (x_c, y_c), by the indicated factor. The parameters for the routine are call expand (right, top, left, bottom, x_c, y_c, factor). The effect produced by the expand subarea operation is similar to a zoom.

Explor An acronym for explicitly provided 2-D patterns, local-neighborhood operations, and randomness, which is a computer language used to build and store graphics and computational functions in a raster-scan format.

fade A controlled effect in which a picture either comes in from black or goes out to black.

FCC The abbreviation for the Federal Communications Commission.

feed reel See **supply reel.**

fidelity The accurate reproduction or duplication of picture and/or sound information.

field One-half—every other line—of the complete picture formed along the television screen's 525 scanning lines. Two interlaced fields form one frame, or a complete picture.

field frequency The rate at which a field is scanned: 30 frames (60 interlaced fields) per second.

field pulse A magnetized impulse encoded at regular intervals along the edge of a length of recorded videotape. It is used as a splicing guide by the videotape editor.

fill A rectangular-subarea operation in the Fortran IV Belfix programming language, used to fill or opaque a specific area with a designated tonal value. The parameters for the routine are call fill (right, top, left, bottom, x, y, n, boundary-n).

film chain A grouping of related equipment for converting motion-picture film into electronic-picture information for videotaping or transmission. The grouping consists of a 35mm projector, 16mm projector, slide projector, and television camera synchronized with the film-projector output. In the TeleCine system the 24-frames-per-second film-projection rate must be converted to the 30-frames-per-second television-transmission rate.

flip-flop A digital circuit that has two states, represented by 0 and 1. A control signal activates the circuit, which remains in one of the positions until the control is reset.

fluorescent screen The target area at one end of the cathode-ray tube. Fluorescent particles in the chemically coated screen glow in direct proportion to the strength of the input signal, thereby reproducing the tonal values of the original subject matter.

Focal A high-level computer language, developed by Digital Equipment Corporation of Maynard, Massachusetts, that is used as a calculating device—either through other programs or by the input of specific commands—for generating timing tapes for additive-color motion-picture-film printers.

format conversion The process of converting and displaying input signals recorded in one format to and in another format, as a TV-video signal to a computer display or a computer-display output to a TV-video signal.

Fortran An acronym for formula translator, an algebraic and logical computer-programming language.

Fortran-coded Explor A simplified version of Explor based on alphamerics, used to teach college-level computer graphics.

Fortran IV Beflix A computer-programming language that combines the area-filling gray-scale capabilities of Beflix with the mathematical capabilities of Fortran.

frame Two interlaced fields, or a complete television picture. The viewer sees 30 frames per second of television transmission; the motion-picture transmission rate, by comparison, is 24 frames per second. See **field** and **field frequency.**

frame frequency The rate at which a full frame of picture information is scanned. The NTSC (National Television Systems Committee) rate is standardized at 30 frames per second.

frame mode An operational mode that displays a preselected film-format outline for image composition. Size and position are under operator control for zoom, crop, and overall image distortion.

frequency The number of vibrations per second or repetitions of sound or light waves in a specific time period. In broadcasting and electronics sound frequency is measured in cycles per second, with the hertz as the unit of measurement.

fringe A Fortran IV Beflix rectangular-subarea operation, used to draw a fringe or outline. The routine is executed by programming the next higher number around the object or area that needs definition. The parameters for the routine are call fringe (right, top, left, bottom, next-to, becomes, flag).

frm The abbreviation for frame in computer-oriented routine-initialization codes.

function See **routine.**

gain The adjustment of contrast in a reproduced picture; "to ride the gain" is a studio expression indicating that console controls are used to make the necessary adjustments.

gemini tape-film system, or **mobile video-film system** A dual-recording technique in which a motion-picture negative and a videotape recording are produced simultaneously. The filming unit consists of a television camera, a motion-picture camera, and a common taking lens.

generation In videotape recording, the duplication of picture information from the original or master tape. The original tape is the **first-generation** print from which the edited master is produced; the edited master is a **second-generation** dupe. **Third-generation** broadcast dupes are made from the edited master. In videotape there is little loss in quality in successive generations.

Genigraphics An image-generation system, developed by General Electric, for the creation, manipulation, digital storage, review, and communication of full-color graphics.

ghost A secondary image, slightly weaker than the original, which results from interference with the transmitted signal.

gray scale The range of gradations or perceptible changes in the black-and-white tonal values of displayed picture information. The purest white obtainable is represented by 0; the deepest black, by 10. The number of intermediate gray-scale values

differs, depending upon the individual system's capabilities.

grid A two-dimensional area in which a packed array of computer-programmed picture elements is stored in a raster-scan representation. The programmer visualizes the two-dimensional surface as a series of squares, each of which is identified by a number. Scanners, called bugs, symbolize the movement of picture elements from one coordinate position to another in compliance with the programmer's instructions. See **bug.**

grid operations The programming, manipulation, and storage of picture elements on a two-dimensional grid. A window, a rectangular programming area within the grid, is depicted in the positive quadrant of the plane, with each square identified by an (x, y) integer pair. The area represented may be changed by the grid-operation routine call window (x-min, y-min)—x-min and y-min become the leftmost column and bottommost row for the new window.

GT40 A system based on a minicomputer and an interactive graphics scope that displays real-time animation on a cathode-ray tube.

hardware The equipment—cameras, computers, monitors, and related modules—needed to generate, process, record, store, retrieve, and display electronic picture information.

head The device on the videotape recorder that scans the tape and records the picture and/or sound information on its iron-oxide surface, which is roughly one-sixth the thickness of a human hair.

helical-scan recording A system of videotape recording in which the picture information is recorded diagonally across the tape in adjacent strips.

high-band recording A videotape recording that eliminates moiré effects and yields a very high signal-to-noise ratio. The picture information is carried over a 10-megacycle band—the standard for 2″ quadruplex broadcasting. See **quadruplex recording.**

high-level languages Computer languages are the means of communication between the programmer and the computer and part of the industry's software. A high-level language (Fortran, for example) is organized in much the same way as a spoken language and written similarly or identically for different computers. Both high-level and low-level, or assembly, languages can be used for animation, and both must be converted to binary form. The translation program employed for low-level languages is short and occupies little memory; for high-level languages, it may occupy a large percentage of the total memory of a minicomputer. A high-level language comprehensible to nonspecialists is used in animated productions for the control-tape-generating system. Some high-level languages have control commands that can be adapted easily to either on-line or off-line computer operations. Compare **low-level languages, machine language,** and **source language.**

hybrid computer systems Special-purpose computer systems combining the extreme accuracy of the digital computer and the high-speed functioning of the analog computer. The digital computer is used for control, storage, timing, and interpolation assignments; the analog computer is used for generating basic shapes, structuring, and animation.

Iconoscope, or **Ike** An RCA television-camera pickup tube that converts visual images into electric signals for subsequent transmission.

image orthicon A sensitive television-camera pickup tube that is especially effective under difficult exposure and lighting conditions. It has generally been replaced by the newer plumbicon tube.

immediate access The instantaneous retrieval and display of previously recorded picture and/or sound information, usually form a roll of videotape.

indirect operation See **off-line operation.**

inlay See **key insert.**

input The recording and storage of picture and/or sound information.

instruction Information, fed to the computer or held in the memory bank, that initiates one or more operations.

interface (1) A set of circuits that links the computer mainframe with compatible external peripherals. (2) The actual connection or means for connecting two related pieces of equipment. See **mainframe** and **peripherals.**

interlaced scanning A picture-scanning process in which the even- and odd-numbered lines are alternately scanned by the electron beam to complete two fields, or a full frame—1/30 second of viewing time. See **field** and **frame.**

interlock The synchronous projection of separate picture and sound tracks, either on film or on videotape.

interpolate In key-frame computer-generated animation, the introduction of intermediate drawings between the animator's extremes. The computer calculates the spacing and position of the inbetween positions from the programmed instructions.

intersync An accessory developed by the Ampex Corporation that synchronizes the output of a number of videotape recorders with live camera sources.

intersync system, or **A & B composite system** An electronic process in which picture information from two separate lengths of videotape is combined to produce a composite that is complete with respect to visual continuity. The system is basically similar to the A-&-B checkerboard-editing system. See **A & B editing, videotape.**

i/o channel An auxiliary processing unit that controls the flow of input and output information between the computer's memory and the peripherals. See **memory.**

ips Inches per second, the unit for measuring the rate at which a length of tape passes over either the record or the playback head.

iron oxide A layer of magnetic particles spread over a thin plastic base, used for recording picture and/or sound information. The magnetic coating is approximately one-sixth the thickness of a human hair.

jack A type of fitting that is used to complete the electrical connection between two related pieces of equipment.

keyboard mode An operational mode used to create image components (in directing data input), font selection, and artwork-record management. The Genigraphics keyboard has a complete set of alphamerics (upper- and lowercase letters, numbers, and special characters) as well as a read key for retrieving and displaying artwork on the console monitor, a write key for storing displayed artwork, an add key for retrieving artwork and adding to the image displayed on the monitor, and a prc key for automatically recording the generated artwork on film.

key-frame computer-generated animation A picture- rather than language-driven electronic-imaging technique analogous to conventional animation key drawings.

key insert, inlay, or **static electronic matte** Similar to the matteing processes used in motion-picture production, this electronically controlled technique is used to insert information from one length of videotape into the background of a second length of videotape.

kilocycle (kc) A unit of frequency equivalent to 1,000 cycles per second.

kinescope recording, telerecording, or **teletranscription** The filming of picture information displayed on a television monitor.

language The means of communication between the programmer and the computer console.

letter A miscellaneous operation in the Fortran IV Beflix programming language, which, along with the type routine, controls the formation of patterns and the display of alphameric characters by numerically encoding the string of entries representing each. The routine is expressed as letter (n, string). See **type.**

light pen An electronic device that resembles a small, pen-shaped flashlight and is used in videotape editing. The editor touches the pen to a display screen that lists the scenes on a roll of previously recorded videotape. The first frame of the scene selected from the menu is immediately visible on a second monitor, giving the editor instantaneous access to any of the scenes on the videotape reel. See **menu.**

lin The abbreviation for line in graphic-oriented routine-initialization codes. It is used to specify vertex descriptions of nongeometric lines.

line A drafting operation in the Fortran IV Beflix programming language. The parameters for the routine are call line $(x_1, y_1, x_2, y_2,$ n, width, frames, speed). This operation draws a straight line by a sequence of horizontal, vertical, and 45° steps from (x_1, y_1) to (x_2, y_2).

linelock In a videotape recorder, the means of control over horizontal and vertical sync signals. This form of electronic registraton makes it possible to combine picture information from several sources and to produce unusual illusory effects.

lines, scanning See **scanning.**

logic In computer programming, the symbols used to define basic concepts.

low-band recording A videotape recording in a low-frequency carrier band (5 megacycles). It has a mediocre signal-to-noise ratio, and moiré effects are present. Compare **high-band recording.**

low-level, or **assembly, languages** Languages that are organized for and suited to a particular computer's operations. Each assembly language is unique and cannot be run on a different computer. Low-level languages are generally used only if the system has a severe memory-storage problem. Compare **high-level languages, machine language,** and **source language.**

machine language The punched cards, control tapes, keyboard operations, or other programming inputs that make a computer system operative. See **data.**

macro, or **macroinstruction** A single computer instruction that stands for a sequence of operations. See **Tarps.**

MAGI An acronym for Mathematical Applications Group, Inc. See **SynthaVision.**

magnetic core A type of high-speed, fast-access computer memory. See **memory.**

magnetic disk See **videodisk.**

magnetic film A film with a magnetic coating of iron-oxide particles over a sprocketed acetate base, used to record motion-picture sound tracks.

mag track A sound track recorded on magnetic film for a motion picture rather than for videotape.

mainframe The computer itself, excluding input, output, or additional storage units. See **peripherals.**

master In videotape recording, an original first-generation print. Derivatives are B master, edited master, and submaster.

megacycle (mc) A unit of frequency equivalent to 1 million cycles per second.

memory, memory bank, or **storage** The computer subsystem used for storing and subsequently retrieving information. Arithmetic operations are generally short-term, fast access memory functions; long-term, slow-access memory systems are used for data and computer-program functions. See **data.**

memory capacity The computer's storage capability, generally measured in kilobits.

menu A list of programming instructions from which an animator may select a particular operating mode. In some ways analogous to a film library, a menu provides immediate access to previously programmed picture data or computer routines. In videotape editing a menu consists of a listing of previously recorded scenes, each of which is identified by a word or number. The editor points a light pen to any of the scenes listed in the display, and the first frame is immediately displayed on a second monitor. See **light pen.**

microcomputer An electronic data-processing device that consists of a number of circuits mounted on a chip of silicon approximately 1/4″ in diameter. The average microcomputer is capable of performing about 100,000 calculations per second, comparable to the operational speed of the minicomputer. The microcomputer is the result of many years of experimentation in miniaturization. Just as the vacuum tube was replaced by the transistor in the late 1950s, the transistor is giving way to large-scale integration. The technique of integrating circuits makes it possible to place thousands of microminiaturized transistors on a chip of silicon no more than a fraction of an inch in diameter.

microphone, or **mike** The device that converts sound into electrical impulses for recording and/or transmission.

microsecond A unit of measurement equivalent to one-millionth of a second.

microwave link, booster signal, or **relay**   An intermediate point at which transmitted signals are received, amplified, and retransmitted. This type of relay system is needed to send sound and picture information to distant areas: the weakened signal must be amplified before it is retransmitted.

minicomputer   An electronic calculator that consists of hard-wired logic systems soldered into rigid patterns on a printer circuit board. The arithmetic functions (operation program instructions) are built into the integrated circuits, which include the brain, or central processing unit (CPU), and the memory circuits that provide the CPU with pertinent data and initiate the programs. The term describes the physical size of the unit rather than its capability.

mix   See **mix** in the editing glossary.

mobile unit   A television studio on wheels—the large van that transports the television cameras and related equipment also contains control consoles for monitoring and transmitting the remote pickups.

mobile video-film system   See **gemini tape-film system.**

mode   A setting or positioning control that activates a specific circuit in electronic equipment, such as record or playback in a sound system.

modulation   The modification of the amplitude, frequency, phase, or intensity of a signal or carrier wave so that it is compatible with other signals that are to be recorded or transmitted.

module   A unit that can be used either by itself or in combination with other units, through appropriate interfacing, to increase the capability of the entire system. In addition to the obvious advantage that a defective part can be replaced with a minimum of downtime, a system can be upgraded constantly by adding more modern units.

monitor   (1) A studio receiver, positioned on the studio stage or housed in a control console, on which the program material is displayed as it is recorded or transmitted. (2) The reviewing process itself.

monochrome   A black-and-white picture.

mosaic   The target area or fluorescent screen at the end of the television-camera pickup tube. Scanned by the electron gun's beam, the fluorescent particles on the screen's surface glow in direct proportion to the magnitude of the input signal, reproducing the tonal values in the original subject matter.

mouse   A light-emitting pencil used for generating artwork and for retrieving scenes during electronic-editing processes.

move   An operation in the Fortran IV Beflix programming language. Before a bug is used, it must be initialized or identified at coordinates by the routine call place (bug, x, y). It can then be moved by the møve routine to another specified x- or y-coordinate and a different square on the grid. The routine is expressed as call møve (bug, to, bug 2). Bug positions cannot be changed by any other subroutine. See **bug.**

move-alter mode   An operational mode that places programmed component parts of the artwork under operator control via rate controls or function pushbuttons.

multilock   An Ampex system for synchronizing sound, recorded on a separate length of magnetic tape, with picture information on a second video-tape.

multiplex   The channeling of several inputs into one composite source.

n   A number written into a programming routine. See **R.**

nanosecond A unit of measurement equivalent to one-billionth of a second.

network A number of television stations in different areas that share facilities and broadcast material.

nixie tube A digital display on a vacuum-type tube. Electrical equivalents are converted into visual information for subsequent readout, and every digit in this type of display is fully formed and independently lit.

noise Ambient sound that seeps into a recording through the equipment itself. See **signal-to-noise ratio.**

NTSC National Television Systems Committee, an association formed in the early 1950s to standardize television-broadcasting procedures. In addition to prescribing controls for transmission bandwidths the NTSC sets the standards for compatibility with respect to the transmission of program material in color and its subsequent reception in black-and-white receivers.

ø In computer programming, the designation for zero. The exception to this statement is the 0 in the Fortran computer languages.

off-line operation A system in which the peripherals are not connected directly with the computer.

on-line operation A system in which the computer and the peripherals that it is designed to operate—an animation stand, for example—are directly connected.

ordinate A programming entry or reference input that defines the specific position of the y-coordinate in the display. An ordinate and an abscissa together are the coordinates.

output The retrieval and display of previously recorded (and stored) picture and/or sound information.

packing density The number of bits assigned to a specific point in a computer-generated display.

paint A rectangular-subarea operation in computer programming that fills every square in the programmed area with a specified number representing one of the gray values or with the output from the random-number generator. The parameters for the routine are call paint (right, top, left, bottom, R_n).

parameter In computer programming, a variable quantity that defines an area within which component parts of an image can be programmed. Restrictions and/or limitations in movement result from the physical capabilities of the equipment itself.

patch A line connection between two related pieces of equipment.

PDP A type of minicomputer used for off-line and on-line animation recording and for generating timing tapes for additive-color motion-picture-film printers.

pedestal (1) A blanking signal. (2) The black level in a televised picture. See **black level** and **clipping.**

pels An acronym for picture elements.

peripherals Auxiliary units connected to a computer that provide additional input, output, and storage facilities. See **mainframe.**

phase A change or shift in the color balance of a displayed picture, analogous to timing in film.

phase-alternation line (PAL) A method of eliminating hue changes during transmission over relatively large distances. The chrominance signal is temporarily delayed at the point of reception by one scanning line and combined with the following signal. In this way the average signal maintained at every line in the display either cancels out or reduces phase errors.

photoelectric cell A light-sensitive device in an electric circuit that transforms variations in light intensity into corresponding electrical impulses.

picture element Any of the component parts of a recorded or transmitted picture.

pixel An acronym for picture elements.

pixlock A mode of operation that provides phasing control of two or more videotape sources through an external reference, assuring horizontal and vertical sync between sources and making possible a variety of effects in the composite output.

place A grid operation in the Fortran IV Beflix programming language. Bugs—base pointers or scanners—symbolize the movement of picture elements from one coordinate position in the grid to another. To introduce a bug, it must be initialized or identified at coordinates by the routine call place (bug, x, y). It can then be moved to a new position by the routine call move (bug, tϕ, bug 2).

playback The process of reconverting electrical signals previously recorded by the electromagnet in the transport mechanism into sound and picture information. The input signal, with its proportional increases and decreases in magnitude, is reproduced as the tape passes over the playback head.

plumbicon A photoconductive television tube, patented by Phillips, with a photosensitive lead-oxide surface.

point-to-point (PTP) area and line mode An operational mode used to create interactive areas and lines, using the cursor to establish vertex locations.

postproduction The editorial work that follows the taping process, including the rerecording or mixing of the dialogue, music, and effects to produce the composite sound track that is synchronized with the accompanying picture.

prc The transfer function in the keyboard operational mode that automatically controls the recording of artwork on film.

preproduction (1) The preliminary preparations, such as concept formulation or scriptwriting, that precede the actual taping process. (2) The assembly of the equipment that is to be used on the set.

print-through The seepage of recorded signals through adjacent layers of magnetic tape wound on a reel.

program A prescribed routine fed to the computer in the form of coded instructions. The programming language, usually industry terminology, activates specific circuits that are designed to perform certain operations.

programmer The person who issues instructions to the computer and initiates the required programs. He may be an animator, producer, director, instructor, or videotape technician.

pulse An amount of voltage or current with measurable duration and magnitude.

quadruplex recording A videotape-recording system in which four rotating heads record crosswise on the moving tape. The head wheel, on which the magnetic heads are mounted, rotates in a plane that is perpendicular to the direction of travel.

quantization The arbitrary assignment of one of two known numbers when the system cannot resolve between them. The assigned numerical quantity is used to solve a problem in which mathematical certainty is lacking in the original equation. Motion-picture laboratories and still photographers, for example, quantize printing exposures. In key-frame computer programming new coordinates or intermediate steps are assigned to some part of an action in order to produce a smoother sequence of motion.

R The symbol for a random number in computer programming. Compare **n.**

ramp generator A module in a video synthesizer that controls the preset speed and length of a programmed sequence of motion or animation.

randnø The symbol for the randøm-number generator in the Fortran IV Beflix programming language.

randøm A miscellaneous operation in the Fortran IV Beflix programming language representing the next output from the randøm generator when restart equals zero. The routine is expressed as randøm (restart). Both randøm and randwd use a computer-stored common table, iR, of randomly selected numbers. The random-number generator, randnø, assigns numerical equivalents representing starting points (ist), skip distances (ipd), and working numbers (nø). The ist and ipd numbers, indexed into the table iR, generate new starting points, skip distances, numbers or parameters. If new ranges or number distributions are required, the entries in the table must be changed accordingly. Compare **randwd.**

random access The retrieval of previously recorded picture and/or sound information from any location within the storage medium—the beginning, middle, or end of a videotape reel, for example.

randwd A miscellaneous operation in the Fortran IV Beflix programming language indicating a randøm word. The function of this routine is to reset the random-number generator when restart equals zero. The routine is expressed as randwd (restart). Compare **randøm.**

raster The tube face on which a television picture is produced by the scanning beam.

readout (1) The display of visual information, as on a monitoring device. See **menu.** (2) The code numbers, referring to specific locations on a roll of videotape, displayed on a nixie tube. See **nixie tube.**

real time The amount of time required to display a sequence of action on a monitoring device, compared to the actual amount of time required to perform the action.

rec The abbreviation for rectangle in graphics-oriented routine-initialization codes. It is used to specify the size, position, and color of the programmed area.

receiver A TV set that can reproduce transmitted picture and sound information.

record To convert sound and/or picture information into a series of electrical signals that vary directly with the magnitude of the input signal. During the recording process the particles of iron oxide impregnated over the tape's plastic backing are rearranged into patterns that conform to the input signal encoded during the tape's passage over the record head in the transport mechanism.

record head In a tape transport mechanism, the device over which the tape passes on its way from the feed to the take-up reel. Preceded by the erase head, which removes previously recorded information, the record head scans the tape and converts the sound and/or picture information into a series of magnetic patterns that conform to the magnitude of the input signal.

rect A drawing operation in the Fortran IV Beflix programming language. The routine draws a rectangle (with number n or a random number specified by R) of indicated width, with right and left the extreme side bounds in which the vertical lines are drawn, and top and bottom the extreme bounds for drawing the horizontal lines. The parameters for the routine are call rect (right, top, left, bottom, ^R_n, width).

rectangular area In computer programming, a subarea within the two-dimensional grid specified by its rightmost column, topmost row, leftmost column, and bottommost row.

reference input The position of the original subject matter displayed to the television camera before the image is logged into the computer for subsequent manipulation.

refresh rate The number of times that a picture is retraced by the electron beam during each second of transmission time. The NTSC standard is 60 fields, or 30 frames, per second.

regen mode An operational mode used to regenerate or refresh the image subsequent to operator adjustments and/or data entry.

register In computer programming, a temporary storage circuit for holding input information in binary language.

relay See **microwave link.**

reliability The adequacy of recording, storage, and display equipment in terms of operational functioning.

remote broadcast A broadcast that does not originate in a television station's home studio.

resolution The amount of detail in a displayed picture. Pictures that are in sharp focus and present details clearly have **high** resolution; soft-focus, blurry pictures that are lacking in detail are **low** in resolution.

retrieve The process of recalling previously stored picture and/or sound information. See **access time, memory,** and **random access.**

r-g-b channels The primary-color channels (red, green, and blue) in the videotape recorder's color system.

routine, or **function** A series of high-level, preprogrammed commands or instructions contained in the computer, which controls specific independent operations or manipulations (drafting-type operations, for example).

rpl An acronym for rectangular plot area in data-plotting routine-initialization codes. It describes a preprogrammed plot area for various plot types such as line, area, and bar.

safety area See **cutoff area.**

saturation The accepted maximum intensity of a particular color or group of colors in a scene.

Scanimate A video-synthesizer computer system that converts an image into electronic signals, modifies the signals, and displays the image on a monitor.

scanning (1) The movement of the electron beam as it traces the picture information on the 525 scanning lines of the fluorescent screen at one end of the television pickup tube (a CRT, for example). (2) The sweep of a record or playback head in a videotape recorder as the tape passes over it.

scanning lines The 525 lines over which the subject matter is reproduced on a monitor and/or receiver. The American standard is 525 lines; in other parts of the world (Russia, Japan, Australasia) the accepted standard is 625 lines.

sec An acronym for circle sector in graphics-oriented routine-initialization codes. It is used to describe the size, position, color, and orientation of a specific circle component.

secam, or **sequential color memory** A French system for controlling color signals during transmission. The sequentially transmitted color-difference signals occupy the same bandwidth. Unlike the method used for correcting phase errors in the PAL system, hue in the reproduced color picture is not affected by phase.

second-generation print A duplicate print made from the original videotape recording. A print made from this duplication would be a third-generation dupe.

selectivity A receiver's ability to reproduce compatible signals and filter out signals transmitted on neighboring wavelengths.

sensitivity The capability of a receiver to reproduce comparatively weak signals with more or less acceptable quality.

sequential color memory See **secam.**

shift A rectangular-subarea operation in computer programming. It moves the entire contents of the rectangular area a specified number of squares in a stated direction. The parameters for the routine are call shift (right, top, left, bottom, up, left, down amount).

```
                                          right
are  call  shift  (right, top, left, bottom, up   ,
                                          left
                                          down
amount).
```

signal (1) An electric pulse that is proportional to the intensity or magnitude of some component part of the picture. (2) The picture itself.

signal-to-noise ratio The ratio, expressed in decibels, between the maximum undistorted signal output and the noise-level output from material that does not carry a signal.

simulation The duplication of a situation or phenomenon that has no physical counterpart.

single-system recording The recording of picture and sound on the same length of film or roll of videotape. The system is used for filming or recording assignments in which the quality of the production is secondary to the speed, such as news coverage. Synchronization between sound and picture is automatic. Compare **double-system recording.**

slant track See **helical-scan recording.**

slicing, or **density analysis** A computerized sampling system developed by NASA in which a negative is scanned by a television camera and relayed through an analyzer, which converts a specified level of density into a selected color. The entire input image is displayed and analyzed on a cathode-ray tube. During the adjustment process that follows, certain colors are dropped and black-and-white controls are reset. The analyzer unit has a number of color channels, each of which contains red, green, and blue color guns and wheels—more than 30 tints for each of the primary colors.

Slo-Mo A disk recording device, which can be used for direct telecasting or electronic editing. Best known for its instant-replay capability, it is also used to create preoptical effects for subsequent use during the electronic-editing process. The unit can freeze a specific frame or slow down or speed up a sequence of action.

smooth A Fortran IV Beflix rectangular-subarea operation used to round off sharp corners or fill in areas that need patching. The parameters for the routine are call smooth (right, top, left, bottom).

software A graphic-operation system, console-operation program, or any utility equipment used to initiate a computer program.

source language Industry terminology translated into a computer-programming language. In order to make a system operative, a source language is often translated into a machine language by means of a compiler.

special-effects generator A device that enables the programmer/editor to add wipes, split screens, transitions, mattes, and other effects. It can be used for color or black-and-white production. Switching is done in the vertical interval.

spot A point of light on the target area, produced by the scanning beam in the television camera's pickup tube. The intensity of the spot varies with the tonal value of the subject matter. It moves continuously over each of the 525 scanning lines to recreate the image 30 times per second.

static computer-generated animation imaging, static computer imaging, or **static display images** A mathematically based electronic system for creating, displaying, and recording images.

static electronic matte See **key insert.**

storage See **memory.**

strobe In computer programming, a system for storing and retrieving input information within a specified period of time.

subroutine A preprogrammed command or series of instructions, contained in the computer, that assigns specific movements to be carried out in conjunction with the primary command or function. Subroutines generally supplement functions or routines. See **drafting operations.**

supply reel, or **feed reel** The reel from which a tape unwinds during the taping, playback, or fast-forward mode before it passes over the record or playback head and winds up on the take-up reel.

sweetening The postproduction enrichment of the original sound track—by the addition of canned laughter or applause, for example.

sync generator A device that provides the synchronizing signal that enables a number of related pieces of equipment to operate in unison.

synchronization (1) The precise alignment of a sound track with the accompanying picture information. (2) The simultaneous operation of two or more related pieces of equipment, as in an interlock.

sync logic The generation of horizontal and vertical blanking and subsequent combination into composite blanking. The phase-controlling pulses that are generated are used to synchronize the horizontal and vertical advance circuitry in television transmission and receiving equipment as well as in videotape recorders.

SynthaVision, or **MAGI** A computer system for generating and animating graphics in which simulated figures are translated into mathematical formulas, displayed as images on a cathode-ray tube, and reproduced on film or videotape.

synthesizer, light-and-sound An electronic device that generates theme music and synchronizes it with a projected picture. The basic process can be traced back to the development in 1962 of a raster-bending video device for the sole purpose of creating background effects for a rock-music group. The unusual colored-light patterns created with the analog ear are now accepted as routine, and programmers create brilliant melodies and original sound patterns instead of duplicating the work of Bach, Brahms, or Beethoven. In operation the electronic device for synchronizing light with sound assigns specific frequencies to certain sounds or to any of the notes that a musical instrument is capable of producing. Lights, colors, and abstract patterns are programmed to respond to the assigned frequencies by producing audio and video effects that would be the envy of any choreographer. Similarly, a frequency analysis of the resonances of human speech can be synchronized with the corresponding screen image. Increases in the amplitude of the input signal in this case produce proportionate increases in contrast.

system A grouping of related pieces of equipment, with each component unit assigned a specific function in the recording, storage, and display of picture and sound information.

tab The abbreviation for tabular in graphics-oriented initialization codes. The routine facilitates the input of columnized data into operator-determined positions, sizes, and colors. Compare **dtx** and **ttl.**

take-up reel The reel on which a tape is wound after it passes over the record or playback head.

tape leader The section of tape that precedes the actual recording. Analogous to the leaders used at the beginning and ending of a motion-picture production, a magnetic leader contains alignment signals and information relative to the production.

tape speed The linear rate of tape travel over the record or playback head in the tape transport.

tape-to-film transfer The reproduction of videotape picture information on motion-picture film. The videotape image is broken down into its red, green, and blue components, and each color is recorded separately. The color separations, enhanced electronically before processing, are combined to produce a composite print.

target The fluorescent screen at one end of the television pickup tube. Scanned by the electron gun's beam, the fluorescent particles glow in direct proportion to the magnitude of the input information.

tarps An acronym for two-dimensional alphameric raster picture system, a computer language developed for generating graphics. It consists of a set of macros written in terms of Beflix. Image-forming textures and tonal values are produced from arrays of closely spaced characters, which are identified by a numerical code.

Telco line A telephone-company line over which picture and sound information is transmitted.

telecast A television broadcast.

telecine The film chain that converts motion-picture film to electronic information for videotape or television transmission. The chain consists of 35mm projectors, 16mm projectors, slide projectors, and television cameras synchronized with the output from the combined projectors. The 24-frames-per-second film-projection rate must be converted to the 30-frames-per-second television-transmission rate.

Telemation An electronic generator that prepares printed graphic material for newsflashes, credits, titles, and other support information.

telerecording See **kinescope recording.**

teletranscription See **kinescope recording.**

television The conversion of light waves into electrical signals, which are transmitted over designated bandwidths to receivers and reconverted into audible-sound and visual-picture information.

test pattern A card that displays lines of varying thickness to a television camera. It is used to check focus and to align equipment before the actual production begins.

thermoplastic recording The recording of sound and picture information on a thermoplastic emulsion. The images in the emulsion layers are processed by heating the film with a radio-frequency current.

time code An identifying code used to measure and locate specific points that are longitudinally removed from the original reference frame. The readout is shown in hours, minutes, and seconds as well as in frames. A time code with a 24-hour capacity recycles itself, showing no difference between A.M. and P.M.

time sharing An arrangement worked out between two producers for sharing the same studio facilities during the period contracted for by one of the principals.

tlit A Fortran IV Beflix rectangular-subarea operation used to transliterate previously programmed numbers. The parameters for the routine are call tlit (right, top, left, bottom, table).

tone A 1,000-cycle sound fed to the audio section of the recording equipment as a reference for checking record and playback levels.

trace A drafting operation in the Fortran IV Beflix programming language, used to trace an arbitrary curve in response to incremental directions and amounts. The programmed directions are expressed by the parameters call trace (x_1, y_1, string, n, width, frames, speed).

transfer The duplication or reproduction of a videotape recording on motion-picture film or vice versa.

transmission The sending of picture and sound information from the studio or point of origin to monitors or home receivers.

transport The mechanism in a videotape recorder that is responsible for moving the tape from the feed reel to the take-up reel (and vice versa) during the record, playback, fast-forward, or reverse mode.

trigger A timing pulse that initiates specific circuit operations.

ttl The abbreviation for title in graphics-oriented routine-initialization codes. It is used to indicate the input or preprogrammed size, position, and color of a title.

TVola A viewing or monitoring device for television, similar to the moviola used by the film editor in postproduction stages.

txt The abbreviation for text in graphics-oriented routine-initialization codes. It is used to control the size position, and color of text material.

type A drafting peration in the Fortran IV Beflix programming language, used to type dynamic alphameric characters. In programming the lower-left corner of the first letter in the string is assigned the location (x_1, y_1). The text, in addition to the letters of the alphabet, may include all of the numbers and symbols that the typewriter keyboard is capable of producing. The parameters for the routine are call type $(x_1, y_1,$ string, n, size, frames, speed).

ultrahigh frequency A frequency over 300 megacycles.

update A miscellaneous operation in the Fortran IV Beflix programming language, used to revise and store previously programmed picture elements on the surface of the two-dimensional grid.

VCR The abbreviation for videotape cassette recorder.

vector (1) A line that graphically represents magnitude and direction. (2) The quantity represented by the line itself.

velocity compensator An electronic accessory that eliminates horizontal distortion (banding) by adjusting or compensating for the scanning rate of the video heads.

vertex (1) The highest point. (2) A specific point at which two lines intersect.

very high frequency A frequency ranging from 30 to 300 megacycles.

video The picture information in a videotape recording or motion picture.

Video Character Generator A computer system that composes alphameric messages, stores them in the memory, and superimposes them on programmed video signals either internally within the unit or externally on a suitably keyed vision mixer.

video compressor A unit that accepts standard composite video signals and performs sampling, digitizing, buffering, and slow-scan TV-conversion operations. It provides outputs for computer use and for the direct connection of slow-scan signals to voice-grade communications circuits. As a television-to-computer interface the device can extract pictorial information from a standard television signal at a rate compatible to that of computers and other digital processing equipment. CVI's model 260 utilizes a sampling technique to obtain a reduced-bandwidth signal from a conventional television signal. With either American 525-/ or European 625-line signals the rate is approximately 16,000 samples per second. Either internal or external frame scan can be used, and data can be taken from either or both of the two interlaced TV fields. A variety of input and output signal options are available. The video compressor incorporates a real-time video output, which allows the programmer to monitor the sampling process. Location of the sampling line is displayed on the TV screen as a series of superimposed white dots, while the wave form produced by the sampled data is also displayed vertically at the left-hand side of the screen. The unit can provide 64 gray levels of video information to a computer or other digital processing equipment.

videodisk A system for recording picture and sound tracks by converting them either into light patterns with a laser beam or into a stream of electrons with a microscope.

video expander A unit that converts slow-scan TV signals to standard TV rates. The CVI model 261 is designed as a companion unit to the video compressor but may be used with any properly formatted analog or digital input signal. Data may be fed into a video expander at a slow rate and used to build up a continuously refreshed image on a standard TV monitor screen. A 12″ magnetic-disk memory rotating at constant speed is utilized for image storage. The image-buildup process may be continuously observed, and there is no fadeout or degradation with time in the resulting display. The unit functions as both a computer-output-display device and as a video communications terminal. In communications applications an incoming slow-scan TV signal is converted to a digital format, reconverted to analog, and recorded on the magnetic-disk memory. Internal circuitry is designed to detect frame-start and line-sync signals, and a choice of two frame rates allows reproduction of low- or medium-resolution imagery.

Videola A videotape-editing system in which edge-numbered 16mm film produced from selected takes is used to make a work print. The frame-by-frame edited work print and a corresponding charting tape are used for the subsequent A-&-B-roll printing that produces the composite videotape.

Video Memory A medium-resolution, four-channel memory device that allows simultaneous recording and playback of four separate synchronized frames of a 12″ magnetic disk, driven at 1,800 rpm by a synchronous motor. Three fixed heads and one movable head with associated write-erase-read electronics are used for data channels, and a fifth head is used for playback of a prerecorded television sync track. Sync, horizontal drive, and vertical drive are available with peripheral equipment. Two writing modes are offered: in the first the operator may choose any desired input-signal modulation; in the second an internal-pulse position-modulation system is used, which allows one dot to be recorded during each TV line, with the input to the modulator determining the location on the line at which the dot falls. An inhibit input to the write amplifier allows additional control over either writing mode. An erase amplifier is utilized so that small segments of the video data may be deleted by means of a light pen or other signal source. The CVI model 404A is intended primarily as a computer-output display device, providing four medium-resolution monochrome channels (250-x-400 elements) or two dot-interlaced 500-x-400-element channels. Synthetic color is readily produced by using three channels on a red-green-blue basis.

video plotter An instrument that converts a wide variety of data inputs into video format and provides continuous storage of this data by means of a rotating magnetic-disk memory. Stored data may be erased completely or selectively or retained indefinitely, even with the power turned off. The output of the CVI model 401A is in the form of conventional industrial TV signals, and data may be displayed on any standard TV monitor. Signals may be videotape-recorded or distributed to remote locations by means of rf (radio-frequency) carrier. Color displays can be achieved with an r-g-b color monitor. The heart of the unit is a 12″ magnetic-disk memory, driven by a synchronous motor at a rate of 1,800 rpm and capable of recording and reproducing information at video bandwidths. A prerecorded TV sync track is reproduced by means of a separate playback head and amplifier, which provide timing information for both internal and external use. An internal-pulse position modulator translates x-, y-, and z-axis input signals into a form acceptable to the disk memory. The plotting accuracy of the unit is 1% or better for full-screen height or width, and an internal electrical grating generator is incorporated for reference purposes, allowing the use of TV monitors with relatively poor geometric linearity.

Video Quantizer An instrument used to achieve unusual variations in the amplitude characteristics of analog-input signals. It is intended primarily for television-studio special-effects production. Its operation involves slicing a normal video signal into eight or sixteen discrete amplitude levels and subsequently mixing the signals to achieve a wide range of visual effects, including keying, tone reduction, outlining, equal-brightness contouring, gray-scale conversion, synthetic-color generation, tinting, and psychedelic patterning (on CVI models 606 and 606A). In synthetic-color generation the various output signals may be mixed and applied to the inputs of a r-g-b color-television monitor or to an NTSC color encoder for subsequent signal distribution or videotape recording.

VideoScene A multiple-camera recording system in which picture information from several sources is combined to produce a single composite tape. The electronic-matteing system makes it possible to combine live performances photographed with one camera with miniature settings, photographs, and other graphic material to produce unusual illusory effects.

video switcher A device in a control-panel console that provides options in the selection of picture information from a number of sources—TV cameras, videotape recorders, or motion-picture projectors, for example.

video synthesizer An analog-scan-conversion system that converts the component parts of an image into electric signals, modifies the signals in compliance with programmed instructions, and displays the continuously changing image on a monitor.

videotape The material upon which picture and sound information are recorded. The tape's recording surface, analogous to the emulsion layer in motion-picture film, consists of a layer of iron-oxide particles over an acetate base. The thickness of the tape, including the sensitive surface and the backing, is measured in mils and is approximately one-sixth the diameter of a human hair. The width of the tape varies from 1/2″ to 2″.

videotape recording An electronic process in which picture and sound elements are recorded on lengths of videotape wound on reels. The versatility of film and the presence of a live telecast can be combined on the tape's iron-oxide surface. Recordings can be played back immediately or at any convenient time thereafter. Unlike film, in which latent images must be developed, videotape needs no processing. During the recording process, the tape unwinds from the supply reel on one side of the transport mechanism, passes over the erase and record heads, and is wound on the take-up reel on the opposite side of the transport mechanism. As the tape passes over the erase head at the rate of 15″ per second, it is degaussed (demagnetized) to remove any residual picture and sound from previous recordings. In the record mode the tape is scanned by magnetic record heads, which rotate at a speed of 14,400 revolutions per minute, or 240 revolutions per second. With 64 scanning lines to the inch, the tape is scanned a total of 960 times per second during its trip from the supply reel to the take-up reel. The sound signal that is recorded by the audio head precedes the corresponding picture by 9 1/4″. The orderly arrangement of the iron-oxide particles on raw tape is altered during the recording process in response to the magnitude of the input signal. Each television field, representing half a frame or 1/60 second, is recorded either vertically or diagonally across the width of the tape, depending upon the method and equipment used: with the helical format the field runs diagonally across the width of the tape; in quadruplex recordings the four rotating record heads place the picture crosswise on the videotape. The picture information from two adjoining fields forms one composite frame that is 1/2″ wide. Each 1/4″ intermeshed field contains 262 1/2 lines—half of the picture information in the 525 scanning lines that make up the completed picture. Two fields are thus equal to one frame, or 1/30 second of viewing time. A standard 12 1/2″ reel of videotape contains 4,800′ of recording surface—enough to hold 115,200 separate images and the viewer's attention for 64 minutes at the standard tape speed. The sound track is recorded in a narrow band along the top edge of the tape; the wide area immediately beneath the sound track carries the picture information. The control track and edit pulses are positioned just below the picture portion; the cue track is located along the bottom edge of the tape. Compare **videodisk.**

videotape sound reader A device used for reproducing sound recorded on magnetic materials. Similar to the sound reader used in motion-picture production, it frees the more expensive videotape recorders for other more profitable uses. The reader is used for identifying as well as for cueing, timing, and analyzing sound tracks.

vidicon tube A television-camera pickup tube with an antimony-trisulfide photosensitive surface. The photoconductive layer is positioned between the transparent signal electrode and the fine-mesh screen that is connected to a cylindrical focusing electrode. At the other end of the tube is the electron gun that generates the image.

Vidifont A TV-display system that produces word messages from a number of different font and size sources in real time.

VLP The abbreviation for Video Long Play, a television disk system developed by N. V. Philips of the Netherlands. See **videodisk.**

voice over An off-camera message or announcement.

VTR The abbreviation for videotape recorder.

wave-form generator A module in a video synthesizer that controls the generation of graphics for subsequent manipulation or animation. It controls sync, frequency, wave shape, and amplitude and frequency modulation.

window A rectangular area represented by the internal two-dimensional programming grid. It is depicted in the positive quadrant of the plane, with each square identified by an (x, y) integer pair. The area represented may be changed by the programming routine call windøw (x-min, y-min)—x-min and y-min thus become the leftmost column and bottommost row for the new window.

word A number of electric or magnetic pulses, generally 12 or 16 bits in a minicomputer, that carry related information and are programmed as a unit.

wow A distortion in reproduced sound caused by variations in the speed of the tape transport mechanism.

wrap The end of a videotape-recording session.

x A coordinate indicating a horizontal point of reference.

y A coordinate indicating a vertical point of reference.

z A coordinate position in a plane other than that used for programming (x, y) coordinates. It indicates the feeling of depth.

zoom A Fortran IV Beflix rectangular-subarea operation used to produce a similar effect to that of a zoom lens. The programmed image is dynamically expanded by the indicated factor, with the expansion centered at (x_c, y_c). The parameters for the routine are call zøøm (right, top, left, bottom, x_c, y_c, factor, frames, speed).

zoom keyer A device that uses matteing patterns on 35mm slides to insert picture information into previously recorded lengths of videotape.

INDEX

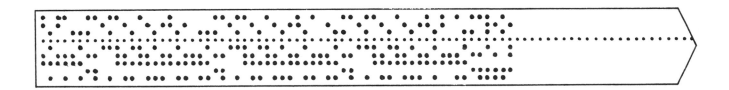

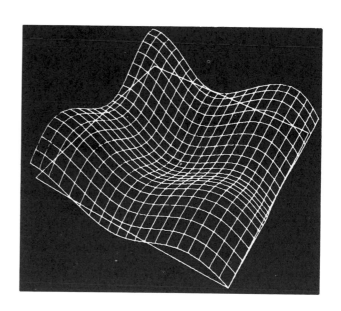

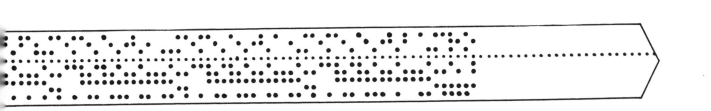